T0074595

Deep Learning in Biomedical and Health Informatics

Emerging Trends in Biomedical Technologies and Health Informatics Series

Series Editors:
Subhendu Kumar Pani
Orissa Engineering College, Bhubaneswar, Orissa, India
Sujata Dash
North Orissa University, Baripada, India
Sunil Vadera
University of Salford, Salford, UK

Everyday Technologies in Healthcare
Chhabi Rani Panigrahi, Bibudhendu Pati, Mamata Rath, Rajkumar Buyya

Biomedical Signal Processing for Healthcare Applications
Varun Bajaj, G R Sinha, Chinmay Chakraborty

Deep Learning in Biomedical and Health Informatics
MA. Jabbar, Ajith Abraham, Onur Dogan, Ana Madureira, Sanju Tiwari

For more information about this series, please visit: https://www.routledge.com/
Emerging-Trends-in-Biomedical-Technologies-and-Health-informatics-series/book-
series/ETBTHI

Deep Learning in Biomedical and Health Informatics

Current Applications and Possibilities

Edited by M.A. Jabbar, Ajith Abraham, Onur Dogan, Ana Madureira and Sanju Tiwari

CRC Press
Taylor & Francis Group
Boca Raton London New York

CRC Press is an imprint of the
Taylor & Francis Group, an **informa** business

First edition published 2022
by CRC Press
6000 Broken Sound Parkway NW, Suite 300, Boca Raton, FL 33487-2742

and by CRC Press
2 Park Square, Milton Park, Abingdon, Oxon, OX14 4RN

CRC Press is an imprint of Taylor & Francis Group, LLC

Library of Congress Cataloging-in-Publication Data
Names: Jabbar, M., editor.
Title: Deep learning in biomedical and health informatics / edited by M. Jabbar, Ajith Abraham, Onur Dogan, Anu Madureira, Sanju Tiwari.
Description: First edition. | Boca Raton, FL : CRC Press, 2022. | Series: Emerging trends in biomedical technologies and health informatics |
Includes bibliographical references and index. | Summary: "This book provides a proficient guide on the relationship between AI and healthcare and how AI is changing all aspects of the health care industry. It also covers how deep learning will help in diagnosis and prediction of disease spread"-- Provided by publisher.
Identifiers: LCCN 2021010428 (print) | LCCN 2021010429 (ebook) | ISBN 9780367726041 (hbk) | ISBN 9780367751548 (pbk) | ISBN 9781003161233 (ebk)
Subjects: LCSH: Diagnostic imaging--Data processing. | Artificial intelligence. | Medical informatics--Medical applications. | Bioinformatics.
Classification: LCC RC78.7.D53 D434 2022 (print) | LCC RC78.7.D53 (ebook) | DDC 616.07/54--dc23
LC record available at https://lccn.loc.gov/2021010428
LC ebook record available at https://lccn.loc.gov/2021010429

ISBN: 978-0-367-75154-8 (pbk)
ISBN: 978-0-367-72604-1 (hbk)
ISBN: 978-1-003-16123-3 (ebk)

DOI: 10.1201/9781003161233

Typeset in Times
by SPi Technologies India Pvt Ltd (Straive)

Contents

Tables

Figures

Abbreviations

ACT	Adaptive Clinical Trials
ADME	Absorption, Distribution, Metabolism, and Excretion
ADMET	Absorption, Distribution, Metabolism, Excretion, and Toxicity
AE	Autoencoder
AHI	Apnea-Hypopnea Index
AI	Artificial Intelligence
AIDS	Acquired Immunodeficiency Syndrome
aLRT	approximate Likelihood Ratio Test
AMP	Antimicrobial Peptide
ANN	Artificial Neural Network
API	Application Program Interface
ASIC	Application Specific Integrated Circuit
ASR	Automatic Speech Recognition
ASV	Adaptive Servo-Ventilation
AZT	Azidothymidine
BBB	Blood–Brain Barrier
BM	Boltzmann Machine
BMU	Best Matching Unit
BO	Bayesian Optimization
CAD	Coronary Artery Disease
CADe	Computer Aided Detection
CADx	Computer-Aided Diagnosis
CAE	Convolution AE
CCTA	Coronary CT Angiography
CHF	Congestive Heart Failure
CIF	Cumulative Incidence Function
CLI	Command Line Interface
CNN	Convolution Neural Network
CNS	Central Nervous System
COPD	Chronic Obstructive Pulmonary Disease
COVID-19	Coronavirus Disease 2019
CPAP	Continuous Positive Airways Pressures
CRF	Circulating Recombinant Forms
CSH	Cause Specific Hazard
CSV	Comma Separated Value
CT	Computer Tomography
CV	Computer Vision
DBN	Deep Belief Network
DCNN	Deep CNN
DITNN	Deep Learning with Instantaneously Trained Neural Networks
DL	Deep Learning
DNA	Deoxyribonucleic Acid

DNN	Deeper Neural Network
DRM	Drug Resistant Mutation
DT	Decision Tree
EC	Eligibility Criteria
ECG	Electrocardiogram
EEG	Electroencephalogram
EHR	Electronic Health Records
EKG	Same as ECG
EliIE	Eligibility Criteria Information Extraction
ELM-LRF	Extreme Learning Machine Local Receptive Fields
ESDA	Exploratory Spatial Data Analysis
FBNN	Feed-Back Neural Network
FDA	Federal Drug Agency
FFNN	Feed-Forward Neural Network
FOG	Freezing of Gait
FSS	Fully Specified Sub-distribution model
FSSAE	Frequential Stacked Sparse Auto-Encoder
GAN	Generative Adversarial Network
GIR	Geographic Information Retrieval
GIS	Geographic Information System
GMDH	Group Method of Data Handling
GPU	Graphical Processing Unit
GRU	Gated Recurrent Unit
HGT	Horizontal Gene Transfer
HIV	Human Immunodeficiency Virus
i2b2	Informatics for Integrating Biology and the Bedside
IE	Information Extraction
IHTSA	International Heart Transplantation Survival Algorithm
IMPACT	Index for Mortality Prediction After Cardiac Transplantation
IPCT	Improved Profuse Clustering Techniques
kNN	k-Nearest Neighbor
LINCS	Library of Integrated Network-Based Cellular Signatures
LSTM	Long Short-Term Memory
LV	Left Ventricular
LYNA	Lymph Node Assistant Model
MBC	Metabolic Breast Cancer
MERS-CoV	Middle East Respiratory Syndrome Coronavirus
ML	Machine Learning
MLP	Multi-Layer Perceptron
MNCH	Maternal, Neonatal and Child Health
MNIST	Modified NIST
mpMRI	Multi-parametric MRI
MRI	Magnetic Resonance Imaging
NB	Naïve Bayes
NCBI	National Center for Biotechnology Information
NCE	New Chemical Entity

NER	Named Entity Recognition
NGS	Next-Generation Sequencing
NIST	National Institute of Standards
NLP	Natural Language Processing
NN	Neural Networks
NSCLC	Non-Small Cell Lung Cancer
OMOP	CDM Observational Medical Outcomes Partnership Common Data Model
ORF	Open Reading Frame
PAD	Peripheral Artery Disease
PeNLP	Preposition-enabled Natural Language Processing
PET	Positron Emission Tomography
PNN	Probabilistic NN
PPMI	Parkinson's Progression Markers Initiative database
RBM	Restricted BM
RCNN	Recurrent CNN
R-CNN	Regions with CNN features
RCT	Randomized Clinical Preliminaries
ReLU	Rectified Linear Unit
RF	Random Forest
RF-LR	Risk Factor Logistic Regression Model
RMSE	Root Mean Square Error
RNA	Ribonucleic Acid
RNN	Recurrent Neural Network
rRT-PCR	real-time Reverse Transcription Polymerase Chain Reaction
RT	Reverse Transcriptase
RT-PCR	Reverse Transcription Polymerase Chain Reaction
SAR	Structure Activity Relationship
SARS	Severe Acute Respiratory Syndrome
SARS-CoV-2	Severe Acute Respiratory Syndrome Corona Virus
SCLC	Small Cell Lung Cancer
SDG	Sustainable Development Goals
SDH	Sub-distribution Hazard
SOFLC	Self-Organizing Fuzzy Logic Classifier
SOM	Self-Organizing Map
SpRL	Spatial Role Labeling
SSA	Sub-Saharan Africa
SSAE	Stacked Sparse Auto-Encoder
SVM	Support Vector Machine
TB	Tuberculosis
TDCS	Time-Dependent Cost-Sensitive
TDNN	Time-Delay NN
UAT	Universal Approximation Theorem
URF	Unique Recombinant Forms
WHO	World Health Organization

Preface

Deep Learning is playing an essential role in biomedical healthcare applications. Deep Learning has recently been widely applied to drug discovery, image classification, neuroscience and molecular computing, disease identification, and pandemics. This edited volume provides different deep learning-based methods for biomedical and health informatics. These methods explore a wide range of applications and possibilities, including disease identification, drug discovery, prediction and monitoring, image analysis and classification.

The book not only covers the state-of-the-art techniques but also is focused on various implementation methods. Each chapter covers all the required methodologies to provide useful information for researchers and beginners. The book has received several chapters from the international community comprising of both practitioners and researchers. Each accepted chapter was evaluated adequately by following very strict evaluation measures.

This book is specially designed for computer science researchers, medical practitioners and professionals working in deep learning, health care informatics& biomedical engineering.

The contents of this book is arranged as nine chapters. Chapter 1 discusses the foundations of deep learning and its applications to health informatics. It covers deep learning history, various models used in deep learning, and deep learning applications in healthcare.

Chapter 2 explains the recent trends in deep learning, challenges and opportunities and provides information about deep learning trends and articles published in healthcare applications.

Chapter 3 reviews the applications of machine learning in the drug discovery process. The authors illustrated how deep learning is applied in the drug discovery process.

Chapter 4 compares in detail deep neural networks for diagnosis of COVID-19. This chapter compares 14 different deep learning models with four classifiers to diagnose COVID-19 by using X-ray images.

Chapter 5 discusses the detection of lung disease with convolution neural networks, which has become almost a standard in image classification.

Chapter 6 surveys deep learning methods to detect COVID-19 cases from radiology images and genome sequences. The main challenges and potential directions of deep learning techniques in combating COVID-19 are also highlighted in this Chapter.

Chapter 7 gives applications of lifetime modeling with competing risks in biomedical science. The breast cancer data set was analyzed with this model using R Software.

Chapter 8 illustrated an extraction and visualization tool for precise detection of maternal, Neonatal and Child Healthcare Geo-locations from Unstructured Data. A parser for extracting geolocations from unstructured MNCH data was also proposed.

Chapter 9 gives details of deep knowledge mining of complete HIV genome sequences in the selected African cohorts. This research implemented a deep knowledge mining framework by combining a batch-learning self-organizing map and a deep neural network classifier

Acknowledgments

We would like to thank all the authors who contributed the chapters to this book series. We would like to acknowledge the hard work by all the reviewers who spent their valuable time in reviewing the chapters. Last but not least, editors would like to acknowledge the love, understanding, and support of family members and colleagues during the preparation of this book. Dr. M.A. Jabbar, Vardhaman College of Engineering, Hyderabad, India; Prof. (Dr.) Ajith Abraham, Machine Intelligence Research Labs (MIR Labs), USA; Dr. Onur Dogan, Izmir Bakircay University, Izmir, Turkey; Dr. Ana Maria Madureira, Instituto Politécnico do Porto Instituto Superior de Engenharia do Porto, Portugal; Dr. Sanju Tiwari, Universidad Autonoma de Tamaulipas, Tamaulipas, Mexico.

Acknowledgments

We would like to thank all the authors who contributed their chapters to this book...

Notes on the Editors

Dr. M.A. Jabbar is a chair, IEEE CS chapter, Hyderabad Section and professor, Vardhaman College of Engineering, India. He obtained his PhD from JNTUH. He has published or presented more than 50 papers in various journals and at conferences. He is a reviewer for Scopus and SCI journals. He served as a technical committee member for more than 50 international conferences. He is the editor of the first ICMLSC 2018 international conference held during 22 and 23 June 2018 at Hyderabad and also the SCOPAR 2019 and NaBic 2019 Conferences. He was awarded the Quarterly Franklin Membership by the London Journals Press.

Ajith Abraham is the director of Machine Intelligence Research Labs (MIR Labs), a not-for-profit Scientific Network for Innovation and Research Excellence connecting industry and academia. The Network, with an HQ in Seattle, USA, has currently more than 1,000 scientific members from over 100 countries. As an investigator/co-investigator, he has won research grants worth over a hundred million dollars from Australia, the USA, the EU, Italy, the Czech Republic, France, Malaysia and China.

Onur Dogan is an assistant professor at Izmir Bakircay University. He graduated from Sakarya University with a bachelor's degree in Industrial Engineering in 2010 and received his master's degree and PhD in Industrial Engineering from Istanbul Technical University in 2013 and 2019, respectively. He studied intelligent decision support systems, lean manufacturing and quality approaches such as QFD, FMEA or DOE during his master's. Process mining was the primary focus of his PhD study. He is one of the first academics to have a PhD on process mining in Turkey.

Ana Madureira obtained her BSc degree in Computer Engineering in 1993 from ISEP, her master's degree in Electrical and Computer Engineering Industrial Informatics in 1996 from FEUP, and her PhD degree in Production and Systems, in 2003, from the University of Minho, Portugal. She became IEEE Senior Member in 2010. She was Chair of the IEEE Portugal Section (2015–2017). She is Teacher Coordinator (Polytechnic Teacher) at the Instituto Politécnico do Porto Instituto Superior de Engenharia do Porto, Portugal.

Sanju Tiwari is a senior researcher. She has worked as a postdoctoral researcher in the Ontology Engineering Group, Universidad Polytecnica De Madrid, Spain. Prior to this, she worked as a research associate for a sponsored research project "An Intelligent Real Time Situation Awareness and Decision Support System for Indian Defence" funded by DRDO in the Department of Computer Applications, National Institute of Technology, Kurukshetra. In this project, she developed and evaluated a Decision Support System for Indian Defence. Her current research interests include Knowledge Based Systems, Intelligent Decision Support, Semantic/Intelligent Query Search Engines, Ontology Integration, Linked Data Generation and Publication, and Knowledge Graphs. She has designed a Smart Health Care Ontology and published it on Linked Data. She has published 21 research papers and two book chapters with international and national publishers.

Contributors

Hasan Arslan
Mathematics Department
Faculty of Science, Erciyes University
Kayseri, Turkey

Hilal Arslan
Department of Computer Engineering
Faculty of Engineering, Izmir Bakircay
    University
Izmir, Turkey

M.B. Bicer
Department of Electrical and Electronic
    Engineering
Izmir Bakircay University
Izmir, Turkey

N. Chandra
Department of Statistics
Ramanujan School of Mathematical
    Sciences, Pondicherry University
Puducherry, India

Onur Dogan
Department of Industrial Engineering
Izmir Bakircay University
Izmir, Turkey

Mercy E. Edoho
Department of Computer Science
University of Uyo
Uyo, Nigeria

Moses Effiong Ekpenyong
Department of Computer Science
University of Uyo
Uyo, Nigeria

Orhan Er
Department of Computer Engineering
İzmir Bakırçay University
Izmir, Turkey

Funebi Francis Ijebu
Department of Computer Science
University of Uyo
Uyo, Nigeria

Devottam Gaurav
Department of Computer Science &
    Engineering
Indian Institute of Technology
Delhi, Inida

O.F. Gurcan
Department of Industrial Engineering
Sivas Cumhuriyet University
Sivas, Turkey

Abdulkadir Hiziroglu
Department of Management
    Information Systems
İzmir Bakırçay University
Izmir, Turkey

Geoffery Joseph
Department of Computer Science
University of Uyo
Uyo, Nigeria

S. Kannadhasan
Department of Electronics and
    Communication Engineering
Cheran College of Engineering
Tamilnadu, India

R. Nagarajan
Department of Electrical and
  Electronics Engineering
Gnanamani College of Technology
Tamilnadu, India

Emre Olmez
Department of Computer Engineering
Yozgat Bozok University
Yozgat, Turkey

Fernando Ortiz Rodriguez
Department of Computer Science
Universidad Automata de Tamaulipas
Tamaulipas Victoria, Mexico

Syed Saba Raoof
Department of Computer science and
  Engineering
Vardhaman College of Engineering
Hyderabad, Telangana

H. Rehman
Department of Statistics
Ramanujan School of Mathematical
  Sciences, Pondicherry University
Puducherry, India

M. Shanmuganantham
Department of Electrical and
  Electronics Engineering
Tamilnadu Government Polytechnic
  College
Tamilnadu, India

Ifiok J. Udo
Department of Computer Science
University of Uyo
Uyo, Nigeria

Kommomo Jacob Usang
Department of Computer Science
University of Uyo
Uyo, Nigeria

Patience Usoro Usip
Department of Computer Science
University of Uyo
Uyo, Nigeria

1 Foundations of Deep Learning and Its Applications to Health Informatics

Syed Saba Raoof, M.A. Jabbar, and Sanju Tiwari
Vardhaman College of Engineering
Universidad Autonoma de Tamaulipas, Mexico

CONTENTS

DOI: 10.1201/9781003161233-1

1.1   INTRODUCTION

Deep Learning (DL) is a sub-domain of Machine Learning (ML). Consequently, ML is a sub-domain of Artificial Intelligence (AI). DL algorithms and approaches emulate the working of human neurons. DL algorithms are implemented by employing Artificial Neural Networks (ANNs); are formed by concatenating numerous neurons, which are called nodes or perceptrons, where all the nodes work as data processing units, the data are transferred from one node to another, and all the nodes are reliable to process the data. Links in-between the nodes are assigned with weights, sums of the weights are computed at every node, and activation functions are applied in order to produce the output; this output of nodes is passed to the preceding nodes. ANNs are categorized into two types based on connection i.e., Feed-Forward (FFNN) and Feed-Back (FBNN). In FFNN a network does not form a cycle, whereas in FBNN a circular connection is formed. Multiple ANNs stacked together produce DL models.

DL plays a vital role in this practical world. Based on the input, the parameters are adjusted and the DL model then fits well on large datasets. DL can be applied in many sectors ranging from banking to medicine.

1.2   HISTORY OF DEEP LEARNING

DL had not emerged overnight; it had evolved steadily over many decades. DL is the outcome of the strong commitment and sheer determination of many researchers. DL had been established in 1943 by Warren McCulloch and Walter Pitts [1] when they implemented a computer-aided system based on the working of biological neurons. Algorithms and mathematical functions were concatenated to emulate the process of neurons. This neuron model was extremely limited, and it didn't have learning capability. Instead, these limitations laid the foundation for both NN (Neural Networks) and DL. Frank Rosenblatt in 1957, [2] developed a Perceptron and demonstrated his work in the paper entitled "The Perceptron: A Perceiving and Recognizing Automaton". The perceptron was the advanced model of the McCulloch-Pitts neuron model that had the learning capability to perform classification. In 1960, [3] the first-ever backpropagation model, developed by Henry J. Kelley, was utilized in the ANN architecture by additional improvements. In 1962, [4] Stuart Dreyfus implemented backpropagation along with a Chain Rule, i.e., the model utilizes a basic differential chain rule rather than using dynamic programming. It had become this one step forward which enhanced the advancement of DL.

Alexey Grigoryevich Ivakhnenko, together with his colleague Valentin Grigor'evich Lapa, in 1965, [5] developed multilayer NN, i.e., the hierarchical depiction of NN, by employing cognitive activation functions, and it had been trained by employing Group Method of Data Handling (GMDH). Alexey Grigoryevich Ivakhnenko is known as the father of DL. In 1969, [6] Marvin Minsky and Seymour Papert issued the book called "Perceptrons", which explained that early perceptron models could not resolve complex functions such as Exclusive OR (XOR). To solve the complex functions, perceptrons have to be arranged within multiple hidden layers so that they adjust the learning mechanisms.

In 1970, Seppo Linnainmaa [7] implemented backpropagation in machine-readable code and published a comprehensive methodology to automate the differentiation process of backpropagation. In 1971, [8] NN had progressed into deeper networks by DL's father. He created an eight-layer Deeper Neural Network (DNN) by employing the GMDH method. Kunihiko Fukushima, in 1980, [9] developed a Convolution Neural Network (CNN) called a Neocognitron, which can identify and distinguish visual patterns, like handwritten characters. John Hopfield, in 1982, [10] developed a Recurrent Neural Network (RNN) known by his name, i.e., Hopfield Network. This network comprises internal memory, thus it can be used for processing sequential data and it even acts as a memory system for addressable content. Paul Werbos, in 1982, [11] proposed the benefits of backpropagation while propagating raised errors while training the NNs. His research work has promoted adapting backpropagation within the research community.

In 1985, [12] the Boltzmann Machine (BM) was developed by David H. Ackley, Geoffrey Hinton, and Terrence Sejnowski, which served similar to RNN and was comprised of only two layers, namely input and hidden layers, and doesn't have an output layer. It is an unsupervised DL model that has only two feasible states, i.e., 0 or 1. Terry Sejnowski, in 1986, [152] created NeTalk, a NN which attempts to learn the pronunciation of English words. Whereas, in the same year, 1986, Geoffrey Hinton, Williams, and Rumelhart, [13] successfully developed backpropagation in a NN. This development in NN initiated the training of complicated DNN with ease.

One further achievement in 1986 [14] was development of a Restricted Boltzmann Machine (RBM). The version of the BM by Paul Smolensky and Yann LeCun, in 1989, employed backpropagation for training the CNN architecture, was deemed as advancement in DL, and even laid the foundation for using Computer Vision (CV) in DL. Universal Approximators Theorem (UAT) was developed in 1989, [15] by George Cybenko. Through his theorem he proved that FFNN with only one hidden layer including a significant number of neurons could estimate and approximate every continuous function and this supplemented the credibility of DL. In 1991, Sepp Hochreiter identified the vanishing gradient problem. In 1997, [16] Sepp Hochreiter and Jürgen Schmidhuber published a paper on LSTM (Long Short-Term Memory). It is a variant of RNN. It led to the DL area being revolutionized and solved the gradient problem. Geoffrey Hinton, Ruslan Salakhutdinov, Osindero, and the, in 2006, [17] developed a Deep Belief Network (DBN) by employing several RBM layers together. This network utilized unsupervised techniques and probabilities to generate outcome of the network and it is widely utilized in areas like image and video recognition, and in capturing motion data. The Graphical Processing Unit (GPU) revolutionary started in 2008 [[18] by N.G. Andrew. A GPU possesses the ability to train DNN considerably faster, even on large amounts of data. Fei-Fei Li, in 2009, [19] the

ImageNet dataset contained 14 million images along with labels and this dataset became exploited as a standard dataset for researchers.

In 2011, there was a struggle for the vanishing gradient. Yoshua Bengio, Antoine Bordes, and Xavier Glorot [20] implemented Rectified Linear Unit (ReLU), an activation function that solves the problem of vanishing gradient. Alex Krizhevsky, in 2012, [21] proposed an AlexNet model, i.e., CNN along with a GPU. This model had won the ImageNet challenge and achieved accuracy of 84%. The blooming of GAN, i.e., Generative Adversarial Network, was initiated in 2014 by Ian Good fellow [22]. GAN started entirely new pathways for DL applications in the fields of art, fashion, and science. In 2016, AlphaGo, [23] a computer algorithm, was developed by Deep Mind technologies that defeats a human mind in the Go game, which is trickier and more complicated than chess. In 2019, Yoshua Bengio, Geoffrey Hinton, and Yann LeCun [24] were rewarded due to their outstanding performance in the advancement of DL and AI. Evolution of DL is shown in Table 1.1.

TABLE 1.1
History of Deep Learning

Year	Evaluation	Reference
1943	A scientific model resembling neurons	[1]
1957	Perceptron	[2]
1960	Backpropagation Model	[3]
1962	Backpropagation with chain rule	[4]
1965	Multilayer NN	[5]
1969	Advanced perceptron	[6]
1970	Computer coded Backpropogation	[7]
1971	Deeper NN, i.e., 8-layered NN	[8]
1980	First CNN	[9]
1982	RNN, named a Hopfield Network	[10]
1982	benefits of backpropagation	[11]
1985	Boltzmann Machine	[12]
1986	NeTalk, a NN	[152]
1986	backpropagation in NN	[13]
1986	Restricted Boltzmann Machine (RBM)	[14]
1989	backpropagation for CNN	[15]
1991	Identified the vanishing gradient problem	[16]
1997	LSTM	[17]
2006	DBN	[25]
2008	GPU	[18]
2009	ImageNet dataset	[19]
2011	ReLU	[20]
2012	AlexNet	[21]
2014	GAN	[22]
2016	AlphaGo	[23]
2019	DL advancement	[24]

1.3   DEEP LEARNING ALGORITHMS

Among all different kinds of DL algorithms, CNN, DNN, and RNN are widely employed algorithms in various fields where DL can be applied.

1. Artificial Neural Network: An artificial neural network (ANN), Neural Network (NN), or simply Neural Net. ANN's are data processing algorithms generated behind the biological neuron structure but on significantly lower scale. NN represents the system of artificial neurons and nodes. In this network all the participating neurons accomplish certain tasks to process the input data and the nodes function as the information carriers; the flow of information from one node to other is generally referred to as activation function. NNs works well even for changeable input data; thus, ANN produces an accurate outcome without the need to remodel the network. NNs are generally arranged in the form of layers composed of certain interlinked nodes, artificial neurons and activation functions. Input data are fed to the network via an input layer, then the processed information of the input layer is passed to one of more hidden layers and the real data processing is done in these hidden layers. These layers are then linked to an output layer.

   Figure 1.1 shows the representation of a NN

2. Feed-Forward Neural Network (FFNN): Deep feed-forward Networks, are known as FFNNs, or Multilayer Perceptrons (MLP). This is a variant of ANN comprised of unidirectional links between nodes. In this network perceptrons are organized in the form of layers, i.e., input, hidden, and output layers. All the perceptrons of one layer are connected to perceptrons of the next layer. Thus, information is always fed forward from one layer to a succeeding layer; hence known as FFNNs. And there will be no connection between perceptrons

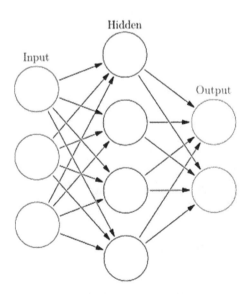

FIGURE 1.1   Representation of a NN [26].

of same layer. The main drawback of this network is that backpropagation is not feasible, thus input function is determined depending on weights. FFNNs are mainly categorized into two types, i.e., Single-Layer FFNNs and Multilayer FFNNs a shown in Figure 1.2.

3. Convolution Neural Network (CNN): CNN is the most widely employed network. It comprises one or many convolution layers that are fully connected in nature. Layers are arranged in 3D form, i.e., width, depth, and height, which are mainly organized for dimensional study, image analysis, and object recognition. CNN is mainly used for image processing tasks and it fits well for spatial data types, like images and videos. The working of CNN relies on the structure of neuron connectivity. Several nodes/perceptrons are employed to ease the task of processing and to produce an accurate outcome. Feature mapping is used to store the features of images and to map to labels. CNN is applied for distinct image processing tasks, like image classification, detection, and segmentation. Architecture of CNN is shown in Figure 1.3.

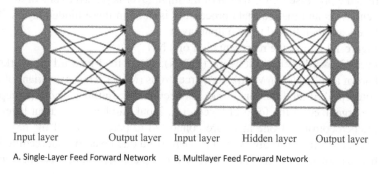

Input layer   Output layer   Input layer   Hidden layer   Output layer

A. Single-Layer Feed Forward Network   B. Multilayer Feed Forward Network

FIGURE 1.2   Architecture of a FFNN ffn [27].

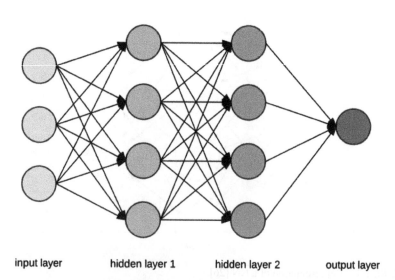

input layer   hidden layer 1   hidden layer 2   output layer

FIGURE 1.3   CNN Architecture [28].

4. Recurrent Neural Network (RNN): It is a FFNN that contains periodic memory loops that fetch data from previous layers. All the nodes in RNN are known as processing units. It is efficient for analyzing data streams. It shares equivalent weight among all the layers thus reduces the parameter complexity. Working is based on the random unit election and it checks the mode of neighboring nodes if it is active then the node computes the sum of weights of the network; if the outcome is positive it switches to active mode and if negative outcome it switches to inactive mode. RNN consist of a specific memory unit that stores the node's information. It is mainly useful and applied where the result relies on preceding computations. It had achieved great success in Natural Language Processing (NLP) applications. RNN Architecture is shown in Figure 1.4.

5. Long Short-Term Memory (LSTM): The LSTM algorithm is a variant of RNN. LSTMs have a chain structure like RNN algorithms, but in LSTM the recurrence unit has a distinct structure: rather than having a single layer, it consists of three or more communicating layers. The key features of LSTMs are cell state, and different gates, i.e., forget gate, input gate, and output gate. LSTMs possess internal devices known as gates which supervise the information transmission. Gates in LSTMs are capable of learning and analyzing which data is important in sequence to preserve and remove. Through this process, they transfer the important information to the chain for prediction. Gates consists of a sigmoid function and pointwise operation. Major applications of LSTM are for text and speech identification, caption generation, and so on. Figure 1.5 represents the LSTM.

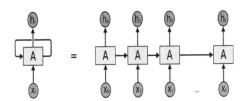

FIGURE 1.4   RNN Architecture [29].

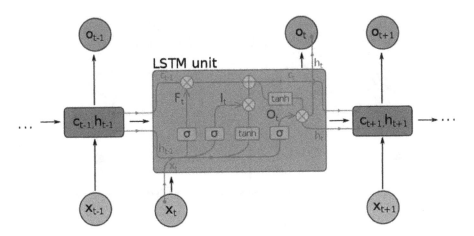

FIGURE 1.5   LSTM representation [30].

6. Autoencoder (AE): this is a specific variant of FFNNs in which both the input and output are the same. The working principle of AE is based on encoding and decoding; the input data is encoded in terms of value given, i.e., input data is compressed after encoding activation is done and then output data is decoded based on its representation. AE mainly comprises three units, i.e., encoder, decoder, and code unit. It is also known as a data compression algorithm; in this algorithm encoding and decoding functions are:
   - Data Specific: Implies that AE can compress/encode similar data to trained data.
   - Automatic Learning: This means, it is quite simple to train specific units that determine to perform on a certain specific input and there is no need of remodeling the network.
   - Lossy: This substitutes that decoded outputs can degenerate when compared with actual input. Thus, it varies with lossless compression. It possesses the proficiency to encode large datasets. AE performs well in dimensionality reduction task and for feature detection. Auto encoder Architecture was shown in Figure 1.6.

7. Boltzmann Machine (BM): Is a stochastic NN algorithm possessing the ability to learn and represent internal depictions and can solve complex problems. All the nodes are interconnected in a circular fashion as represented in the following figure. It is known as a stochastic model as it is generative in nature, i.e., it

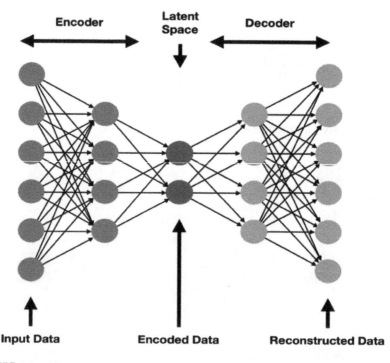

FIGURE 1.6   AE Architecture [31].

generates distinct parameters for distinct models rather than utilizing fixed parameters. It fits best for specific data and develops a binary reference system. Figure 1.7 represents a Boltzmann machine.

8. Generative Adversarial Network (GAN): is a semi-supervised or un-supervised model. It comprises of two NN parts, i.e., generator and discriminator.
   - Generator: A NN, reliable to generate new data similar to actual data as per the given domain.
   - Discriminator: Responsible for classification task which classifies the system generated data from actual data.

   Adversarial GAN term implies that there exists competition between generator and discriminator. It is most broadly implemented in CV and NLP applications. GAN Architecture is shown in Figure 1.8.

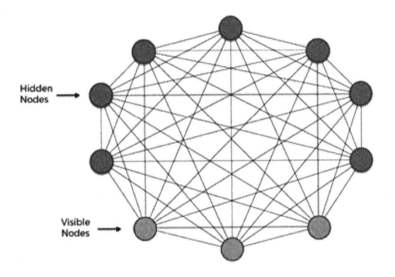

FIGURE 1.7 Boltzmann Machine (BM) [32].

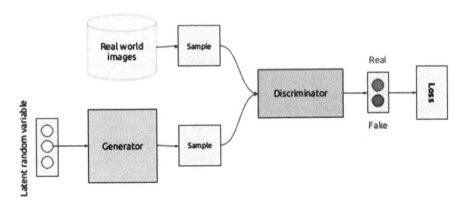

FIGURE 1.8 GAN Architecture [33].

1.4 APPLICATIONS OF DL IN HEALTHCARE

In the view of beneficial outcomes of highly efficient engineering, a torrent of health-care data had been achieved in past decades, i.e., data like medical imaging, public health, drug discovery, proteomic analysis, and so on. Certain DL applications in health informatics are discussed here and Table 1.2 demonstrates the summary of DL applications in health informatics.

1.4.1 MEDICAL IMAGE PROCESSING

ML in the medical imaging field is an effective implementation but it involves human power, i.e., a proficient physician. This drawback had been overcome by DL techniques. Amongst the greatest achieved DL application globally is image analy-sis. DL approaches had recorded the best accurate results in the fields of object recognition, image processing, segmentation, and classification. DL algorithms play a vital role in the medical field in recognizing diseases depending on the input images, thus forming medical imaging a distinct application for DL. These algo-rithms can be applied for detecting, recognizing, and diagnosing different diseases and can even be applied for classificationby training the model as appropriate, and it had recorded high results in this field. Table 1.2 summarizes the applications developed by different researchers.

1.4.2 DRUG DISCOVERY

Drug discovery and drug analysis will play forever a significant role of drug investi-gation globally. DL can be implemented in this sector of drug discovery to ease the role of drug researchers as it is a complicated task and DL saves time, resources, and expenditure. Even though many traditional solutions have been designed, they do not meet all the challenges of drug discovery. This constraint can be met by employing deeper algorithms like DNN, RNN, AE, and so on to detect the characteristics of drugs, drug reactions, and chemical design.

1.4.3 PROTEIN STRUCTURE ANALYSIS

A chain of amino acids describes the protein structure. The main challenge in this sector is detecting stereoscopic structure from the linear structure, as an inappropri-ate structure induces a broad range of diseases instead of curing a particular disease. Since DL can even depict complex data in hierarchal form, DL can be implemented in this sector for detecting and predicting the stereoscopic structures and it is effec-tive in analyzing protein structures. J. Lyons et al. [99], developed a DL-based model to predict the sequence of structural angles. M. Spencer et al. [100] devel-oped an approach to predict the stereoscopic structure. Li et al. [89], implemented a CNN model for detecting stereoscopic protein structure and recorded 82.6% accuracy.

TABLE 1.2
DL Models Used for Various Applications of Healthcare

Domain	Application	Method employed	Reference
Medical image processing	Tumor detection	DNN, RNN	[34–37]
	Brain tumor segmentation, classification, brain image construction	DNN, CNN. AE	[38–42]
	Epileptic seizures prediction based on EEG, Epileptic seizures prediction, EEG patterns classification EEG lapse detection, and Anomaly detection based on EEG	DBN, RNN	[43–47]
	Segmenting organ Segmentation of Prostate MR Hippocampus segmentation Brain image segmentation Segmentation of glands Tissue segmentation from CT scans Pancreas segmentation Left ventricle segmentation Segmenting knee cartilage	CNN, DBN, DNN, AE, SAE	[36,38,48–61]
	Classification of neuron cells Brain cell classification Classifying tissues Alzheimer diagnosis	CNN, DBN, Convolutional DBN	[42,62–70]
Bioinformatics	Drug discovery Gene classification Protein structure interaction RNA protein binding	DBN, DNN, AE	[71–76]
Medical Informatics	Disease prediction Medical data monitoring, mining Monitoring human lifestyle	AE, CNN, DBN, DNN, RNN	[77–85]
Protein Structure	Secondary structure prediction Protein link prediction Protein region prediction Protein disorder prediction	DNN, RNN, SAE, CNN	[86–98]

1.4.4 BIOMARKERS

Biomarkers play a key role while accessing medical output, for disease identification and monitoring. The major challenge in this sector is detecting accurate and advanced drugs for particular biomarkers. A. Zhavoronko et al. [101], implemented a DNN and CNN based model for biomarker and drug discovery. Eugene B et al. [102], proposed a system to photo age clock a biomarker for detecting age. M.S. Pepe et al. [103], implemented a biomarker for detecting cancer at an early stage.

1.4.5   BIOINFORMATICS

Traditionally, bioinformatics algorithms basic domains are detection, diagnosis, and prevention of diseases which are subjected by genomics, epigenomics, and pharmacogenomics. The key challenge is to overcome transcriptional errors of chromosomes, cell division, DNA representation, cancer or disease diagnosing. DL algorithms meet these challenges by employing different deeper algorithms like DNN, Recurrent CNN (RCNN), RNN, etc., which can learn, analyze, and detect the cytogenetic complex structures and their working, which can help to diagnose the various diseases.

1.4.6   MEDICINAL INFORMATICS

Medical informatics is the study of enormous medical clustered data in an attempt to segregate data that helps to make decisions accurately, i.e., to build an accurate Decision Support System (DSS) in the medical domain that further also enhances the accessibility of medical data. The traditional Electronic Health Record (EHR) system provides access to all the medical data but poses some drawbacks like data complexity and data irregularity in disease, data that will lead to learning complexity. Since DL apprehends supervised data and unsupervised data and, as it learns and trains the model based on the unique structure of data, thus DL performs well on large datasets. Furthermore, DNNs are capable to group multiple data aspects, thus it can manage various models' data accurately. Therefore, DL is universally recognized and implemented globally in this field [104]. R. Mioto et al. [78], developed an AE and DL model named DeepPatient to acquire patients' data from EHR without manual effort and the proposed model also predicts the prospective patients from EHR data. S. Mehrabi et al. [80], proposed temporal patterns and codes to detect and diagnose the diseases based on DL. H. Shin et al. [105], implemented a CNN model to inter-relate the data and images on large radiology datasets. J. Fatuma et al. [85], compared several existing models for predicting the hospital /readmissions from EHR data.

1.4.7   PUBLIC HEALTH

Public Health is the study of the outbreak or pandemic situations and its applications, like pandemic surveillance, air pollution check, drug safety, and stimulating disease behavior. Data processing and computerized approaches assist to create models that serve in this field. DL methods work well as they support large data-based models and generate best results. Garimella et al. [106], developed an application to predict the lifestyle of diseases. B. Zou et al. [107], proposed a DL model to predict three categories of intestinal diseases. B. Ong et al. [108], implemented a system to forecast the level of pollution in the air. DL models used for various healthcare applications are shown in Table 1.2.

1.5 CASE STUDIES

1.5.1 DEEP LEARNING IN LUNG CANCER PREDICTION

Disease where body cells multiply uncontrollably is known as cancer.

Lung Cancer: Cancer that initiates in the lung region is called lung cancer. Lung cancer can extend to lymph nodes and even to other body parts. Similarly, cancer from the rest of the body may even spread to the lungs; this is known as metastases. Lung cancer is categorized into two types, i.e., Small Cell Lung Cancer (SCLC) and Non-Small Cell Lung Cancer (NSCLC).

Lung cancer ranked the highest projected incidence and recorded the highest mortality rate among all types of cancer globally in both men and women, and the lowest survival rate. Therefore, lung cancer is the most significant cancer concerned worldwide [109].

1.5.1.1 History of Deep Learning in Lung Cancer Prediction

Stojan Trajanovski et al. [110], proposed two-tier DL architecture for detecting lung cancer risk analysis. In the first phase, nodules are identified by a nodule detector. In the second phase, risk of cancer is identified by employing deeper NN. Experimental analysis is carried out on three different datasets, i.e., the National Lung Screening Trial (NLST), Kaggle, and the Lady Hardinge Medical College (LHMC), with a recorded accuracy of about 86-94% for all the three datasets. Diego Ardila et al. [111], developed a DL model that employs past and present patient CT scans and recorded 94.4% accuracy. R.R. Subramanian et al. [112], utilized pre-trained models via VGG-16, AlexNet, and LeNet; achieved around 97-99% accuracy, and this model had served as a persistent and supportable diagnosing model for detecting lung nodules. Furthermore, python-Tkinter a user-friendly interface had been developed so that laypersons can also use it with ease. N. Coudray et al. [113], trained a DCNN model, i.e., Inception v3 on pathology images along with higher accuracy of 97.0%. Yiwen Xu et al. [114], implemented DL models by employing time series; the model was trained and validated on two datasets consisting of 179 and 89 patients withIII stage NSCLC, consisting of 581 and 178 scans respectively and achieved an accuracy of around 74%. P.M. Shakeel et al. [115], developed a model based on Deep Learning with Instantaneously Trained Neural Networks (DITNN) and Improved Profuse Clustering Techniques (IPCT). A CIA dataset was utilized to perform experimental analysis and achieved 98.42% accuracy. Nasrullah et al. [116], proposed a Region Based Convolutional Neural Network (R-CNN) and UNet model along with MixNet and tested the model on the LIDC-IDRI dataset and achieved 99.0% accuracy. Michael T. Lu et al. [117], developed a Convolutional Neural Network – Lung Cancer (CXR-LC) model for detecting the risk by utilizing EMR data and recorded 74% accuracy.

1.5.2 DEEP LEARNING IN BREAST CANCER PREDICTION

Breast cancer: Cancer initiating in the breast region is called breast cancer. Generally, tumors are formed in the breast which can be identified by x-ray, mammograms, breast MRI, and ultrasound. The most used tests are mammogram and

MRI. It usually occurs in women but men are also affected by it [109]. It is one of the dangerous diseases, and it is the major cause of cancer demise among women globally. Breast cancer treatment is categorized into two types, i.e., local treatment and systematic treatment. Radiation therapy and surgery come under local treatment, and hormone and chemotherapy fall under systematic treatment. Generally, to obtain accurate results a combination of both treatments is employed. Even though it is the second leading cancer caused among women, the survivability is high along and early prediction may save up to 97% of patients for about 5 years or even more.

1.5.2.1 History of Deep Learning in Breast Cancer Prediction

Yun Liu [118], developed a Lymph Node Assistant model (LYNA) for detecting Metastatic Breast Cancer (MBC). Camelyon16 challenge dataset, and DS2 (separate dataset) were used for evaluating the performance of the model and recorded an area under the curve (AUC) of 99% and 99.6% respectively, and even developed a framework for pathologists to access and use the algorithms efficiently. A. Yala et al. [119], implemented three DL models, i.e., RF-LR (Risk Factor Logistic Regression model), ResNet18, and a hybrid model that is the combination of preceding two models based on mammography, which was self-collected from the medical center and recorded an accuracy of 95%. L. Shen et al. [120], developed a deeper CNN model and compared with two DL models, i.e., VGG and ResNet. The Curated Breast Imaging Subset of DDSM (CBIS-DDSM) dataset was used for experimental analysis and obtained 88% accuracy and the model is publically available at https://github.com/lishen/end2end-all-conv. S.U. Khan et al. [121] proposed a DL framework based on transfer learning and CNN models, i.e., GoogLeNet, ResNet, and VGGNet.

Two datasets are utilized for analyzing the performance of the model; a standard benchmark dataset and the other is collected from the Lady Reading Hospital (LRH) and recorded 97.52% accuracy. J. Xie et al. [122], proposed an AE model by using features of Inception v3, and Inception_ResNet_V2 models based on histopathological images BreaKHis dataset was used. L.Q. Zhou et al. [123], had utilized three pre-trained CNN architectures, i.e., ResNet-101, Inception-ResNet V2, and Inception v3 were evaluated on a dataset collected from Tongji Hospital and obtained accuracy between 78% and 82%. Q. Hu et al. [124] proposed a deep Computer-Aided Diagnosis (CADx) model, using transfer learning and Multiparametric Magnetic Resonance Imaging (mpMRI) images are used for performance analysis. Bryan He et al. [125] implemented a DL model for detecting the local gene expression from histopathology images a new dataset employed; available at [24] and [126].

1.5.3 DEEP LEARNING IN HEART DISEASE PREDICTION

Heart Disease: The term heart disease usually referred to as cardiovascular disease. It is the condition when blood vessels are blocked or narrowed that results in heart stroke or chest pain, and even may affect heart valves, or muscles. Men and women may encounter different symptoms. For example, men may suffer through chest pain, whereas women may show many symptoms along with chest pain, such as difficulty in breathing, extreme fatigue, and nausea. Family history, diabetes, obesity, high blood pressure, and stress are some of the common causes of heart diseases [109]. Atherosclerosis, arrhythmia, Coronary Artery Disease (CAD), cardiomyopathy, congenital heart defects, and heart infections are the different types of heart diseases. Electrocardiogram (ECG or EKG), carotid ultrasound, echocardiogram, CT scan, x-ray, heart MRI are some of the tests to detect heart diseases. Invasive procedures and surgeries are the treatments [127].

1.5.3.1 History of Deep Learning in Heart Disease Prediction

D. Medvid et al. [128], developed a DL model for heart transplantation, namely International Heart Transplantation Survival Algorithm (IHTSA) and Index for Mortality Prediction After Cardiac Transplantation (IMPACT), to determine the survival rate after transplantation. Data were collected from the United Network for Organ Sharing (UNOS) registry which comprises a dataset of 27,860 transplantations, of 27,705 patients and recorded 65.4% and 60.8% ROC values for IHTSA and IMPACT model, respectively. R. Poplin et al. [129] developed DL models; a Classification model to predict binary risk factor, a Regression model to predict continuous risk factor, and a MACE model. The UK Biobank and EyePACS-2 K datasets were utilized for performance analysis of the model and recorded 97.5% of AUC. M. Zerik et al. [130], developed a DL model based on Multi-scale CNN and Convolutional AE (CAE) to segment LV myocardium and to label LV myocardium, respectively. Analysis was done on Coronary CT Angiography (CCTA) scans. Performance of the model was measured by employing Dice-coefficient and 10-fold cross-validation and recorded 91% and 80% respectively. F. Ali et al. [131], implemented a Smart Healthcare System based on Feature Fusion and Ensemble DL techniques to predict heart disease. Sensor data and EMR data is used for experimental analysis and recorded 98.5% accuracy. S. Khade et al. [132], implemented a DL model to predict the Congestive Heart Failure (CHF) and category of failure and even detect the heart failure severity, and recorded accuracy of 84-88.30%. K.H. Miao et al. [133], proposed a DNN system comprising of two models: one for classification and the other for diagnosing, i.e., a prediction model. The dataset was collected from Cleveland Clinic and recorded 83.67% accuracy.

1.5.4 DEEP LEARNING IN BRAIN TUMOR PREDICTION

Brain Tumor: or brain cancer is a nodule or mass that grow abnormally in the brain region. It is categorized into different types based on the stage, such as noncancerous tumor or benign tumor, cancerous tumor or malignant tumor. It can initiate in the brain itself, known as primary tumor, or can spread from other organs to brain, known as metastatic or secondary tumor. The rate of growth and its location specifies the affect to medullar nervous system working. Brain metastatic, glioma, acoustic neuroma, ependymoma, medulloblastoma, pineoblaatoma, etc. are some types of brain tumor. Symptoms depend on the size of tumor, and location. CT, PET, or MRI scans are used to detect the tumor. Based on the type, size and tumor location treatment is carried out; generally, biopsy is the most widely adapted treatment to cure brain tumor [109].

1.5.4.1 History of Deep Learning in Brain Tumor Prediction

H. Mohsen et al. [134], developed a DNN model to classify the brain tumor into four types, i.e., benign, metastatic, sarcoma, and glioblastoma. Fuzzy C- means, Diffusion-weighted imaging (DWT), Principal Component Analysis (PCA), and DNN techniques are used for segmentation, feature extraction, and classification, respectively. For experimental analysis, a dataset was collected from Harvard Medical School which consists of 66 MRI's where 22 were noncancerous and 44 were cancerous images and recorded classification rated in between 86.36% and 96.97% for different employed algorithms. A. Ari et al. [135], developed a CAD system to detect and automatically classify the tumor. The system was divided into three stages, i.e., image preprocessing, segmentation, and classification is done based on ELM-LRF (Extreme Learning Machine Local Receptive Fields). A 16-patient dataset was employed for experimental analysis and obtained 97.18% classification accuracy. Z. Sobhaninia et al. [136], developed LinkNet, a DL architecture for a segmentation brain tumor. A CE-MRI dataset used for experimental analysis recorded 73% and 79% Dice-coefficient for the single and multi-layer networks. M.A. Khan et al. [137], implemented a Fully Automated Multimodal DL model to classify the tumor. BraTs-2015, 2017, 2018 datasets were utilized for experimental analysis and accomplished 98.16%, 97.26%, and 93.40% classification accuracy respectively for all three datasets. T. Saba et al. [138], proposed a DL and transfer learning model to detect tumors. Segmentation had been done by employing the Grab Cut approach, VGG19 for feature extraction, data fusion, and deep learning approaches are also applied for feature extraction and collection, and DT, KNN, SVM, LGR, and LDA methods are used for classification. The model was trained on different datasets, i.e., BraTs 2015–17 and recorded 98.78%, 99.63%, and 99.67% tumor detection accuracy, respectively.

1.5.5 DEEP LEARNING IN PARKINSON'S DISEASE PREDICTION

Parkinson'ss disease: central nervous system or brain disorder that affects the body motion leading to the quivering, slowing of the moment, unbalancing, and stiffness. Symptoms are not evident at the early stages or slight symptoms may be apparent like talk becoming weak or mumbled, arms will not swing when walking, and at later stages symptoms worsen. A patient at later stages suffers from trouble while talking and walking and even encounters sleep and mental problems like memory loss, depression, fatigue, etc. The major cause for this disease is that brain neurons which control the motion of the body die or become impaired. Diagnosis is based on neural examination and the patient's medical history [139]. Treatment is not yet available for this disease but some medications, or surgical treatments can give relief to the patient [109].

1.5.5.1 History of Deep Learning in Parkinson's Disease Prediction

S.L. Oh et al. [140], developed a model to detect the Freezing of Gait (FOG), a Parkinson's disease using DL techniques. Analysis was done on a 21-patient dataset and obtained a geometric mean of 90%. Rodr'ıguez et al. [141], implemented models to detect FOG. J. Wingate et al. [142], developed a novel DL architecture to detect Parkinson's disease by employing DNN and RNN algorithms. MRI scans and dopamine transporters scans were utilized for experimental analysis and recorded 85-% accuracy. R. M. Sadek et al. [143], developed an ANN model and a dataset was collected from Oxford University which comprised of voice recordings of 31 people among which 23 were patients. Total recordings were 195 and the detection rate of system recorded 100% accuracy. J.M. Tracy et al. [144], developed a deep phenol-typing model to detect Parkinson's disease at an initial stage by utilizing voice recordings collected from the Synapse research portal and recorded 85% accuracy. W. Wang et al. [145], novel DL approaches had been developed and compared with traditional ML and Ensemble Learning techniques on a dataset collected from Parkinson's Progression Markers Initiative database (PPMI), and 96.45% of detection accuracy was recorded.

1.5.6   DEEP LEARNING IN SLEEP APNEA PREDICTION

Sleep Apnea: is a sleep disorder wherein breathing stops and starts frequently. The major symptom of sleep apnea is when one sleeps snoring all night and still feels tiredness, then it may be due to sleep apnea. Three main categories of sleep apnea are: complex sleep apnea syndrome, central sleep apnea, and obstructive sleep apnea. Symptoms include loud snoring, insomnia, hypersomnia, irritability, air gasping, and morning headaches. Causes are related to family history, obesity, narrow airway, neck problems, smoking, diabetes, heart, liver, and nasal problems. Nocturnal polysomnography tests and home sleep tests are used to detect sleep apnea. Continuous Positive Airway Pressures (CPAP), supplement oxygen, Adaptive Servo-Ventilation (ASV) are the treatments majorly adapted globally [109].

1.5.6.1   History of Deep Learning in Sleep Apnea Prediction

T. Van et al. [146], proposed a DL algorithm that automatically extracts features and predicts the disease, and LSTM is employed for classification. The sleep-Heart-Health-Study1 dataset is used for experimental analysis and obtained 99.9% accuracy. M. Hafezi et al. [147], developed a portable sleep apnea monitoring device based on accelerometer and DL techniques via CNN, RNN, LSTM, and hybrid DL model combination of CNN and LSTM. An Apnea-Hypopnea Index (AHI) is used for evaluating the model and achieved 84% accuracy. K. Feng et al. [148], proposed a sleep apnea model based on Frequential Stacked Sparse Auto-Encoder (FSSAE) and Time-Dependent Cost-Sensitive (TDCS). The PhysioNet dataset was utilized for experimental analysis and accomplished an accuracy of 85.1%. Li et al. [149], implemented a sleep apnea model for classification by employing a decision fusion approach and Stacked Sparse AutoEncoder (SSAE). DL methods used in disease prediction was presented in Table 1.3.

TABLE 1.3

DL Methods Used in Disease Prediction

Case Study	Methods	Reference
Lung cancer	CNN, RNN, Deeper NN, DCNN, AlexNet, LeNet, UNet, and MixNet	[110–117]
Breast cancer	CNN, RNN, DNN, AE, VGG, ResNet, Inception, GoogleNet, RF-LR	[118–125]
Heart diseases	Convolutional AE, Feture Fusion, Ensemble DL, Regression models, DNN	[128–133]
Brain tumor	DNN, DWT, PCA, Fuzzy, ELM-LRF, LinkNet, transfer Learning, VGG19, DT, (K Nearest Neighbor (KNN), Support Vector Machine (SVM), Local Ridge Regressor (LRR), Linear Discriminant Analysis (LDA)	[134–138]
Parkinson's	DNN, RNN, Novel DL architectures, EL, ANN	[140–145]
Sleep Apnea	FSSAE, SSAE, CNN, RNN, LSTM	[146–149]

1.6   DEEP LEARNING CHALLENGES IN HEALTH INFORMATICS

DL compared with ML algorithms confront various challenges even though it had demonstrated superiority in image processing tasks in the healthcare sector like segmentation, recognition, feature extraction, feature selection, and classification. Clinical data is extremely complicated; thus, it cannot be interpreted by human beings easily, and it cannot be interpreted by standard quality DL algorithms. For example, DNN is insufficient to detect biological data relationships [104]. High accuracy in DL can be achieved by training the model on a large dataset and this is not possible in most cases, thus this leads to the over-fitting problem. Furthermore, DL algorithms are inappropriate for all sorts of diseases. The following aspects explain some of the challenges encountered by DL (Figure 1.9).

1. Data Volume: The basic core of DL algorithms to achieve high accuracy is to contain an enormous volume of data. In the healthcare area, a large number of people won't have access to healthcare. Thus, we cannot procure enough patient data as we need for training the immense DL model. Perception of type of diseases is the most difficult and complex tasks. The volume of healthcare data needed for training an efficient DL model will be far greater compared to data required in other fields (like text or speech classification, media classification, and so on).

2. Data Quality: Medical data is usually known as dirty and noisy data, i.e., it is in the following form: unstructured, heterogeneous, noisy, incomplete, and ambiguous. Training an efficient DL model on such large and diversified data is a challenging task. It is necessary to examine various problems like data redundancy, missing values, and data scarcity. Preprocessing of data is a challenging task and often it is a time-taking process.

3. Data Variability: Global population is about 7.8 billion, thus acquiring broad and apathetic volatile data is an exceedingly difficult task.

FIGURE 1.9 Challenges of DL in healthcare.

4. Temporality: Diseases will always be advancing and changing with time in a non-descriptive manner. Existing DL systems are not capable of working with changing time in natural occurrence. Developing models that can work well with transient and secular data requires new alternative and innovative approaches.

5. Uncertainty: Constantly diseases will be originating in an obscure way. Hence, developing models that can fit this uncertain data is a crucial task.

6. Interpretability: Interpretability by many DL researchers is considered a major challenge in the healthcare area even though these algorithms had achieved desirable outcomes. DL models are frequently considered as black boxes in healthcare as it is difficult to elucidate where and how the DL models play well, as the outcome of medical systems is actively engaged with life and death issues. Still, it is a difficult task to convince physicians regarding what exactly the applications endorse individuals to perform.

7. Complexity: Challenges in the healthcare domain are more complex when compared to other domains like visual recognition, language translation, speech recognition, etc. because diseases are extremely heterogeneous and, for the majority of diseases, full information doesn't exist.

8. Causal Inference: Detecting the causal connection among diseases and their respective treatments is a difficult task. In [150] implementing a DNN model to address casual inference challenge through studying the hidden relationship features.

9. Legal and Ethical Issues: Privacy of patient data uprising challenges in the healthcare sector [110]. In most nation, patient's records are considered confidential data and shared only among medical organizations, not across other organizations. Therefore, the proprietor of this information generally should formulate terms and conditions to carry out the research. Therefore, such restrictions lead to limited data analysis and even produce low performing and inefficient results.

1.7   CONCLUSIONS

This chapter summarizes the basic concepts and models in deep learning. We reviewed the deep learning models such as RNN, CNN, GAN, LSTM, and AE. DL models have the power to mine knowledge from biological and biomedical data. We also provided detailed case studies of deep learning in healthcare. Challenges faced in applying DL in healthcare were also discussed in detail. We believe that this chapter will help in the further development of DL in the healthcare domain.

REFERENCES

[1] McCulloch, W.S. & Pitts, W. (1943). A logical calculus of the ideas immanent in nervous activity. *The Bulletin of Mathematical Biophysics*, 5(4), 115–133. doi:10.1007/bf02478259

[2] Rosenblatt, F. 1957. The perceptron, a perceiving and recognizing automaton Project Para. Cornell Aeronautical Laboratory

[3] Kelley, H.J. (1960). Gradient theory of optimal flight paths. *ARS Journal*, 30(10), 947–954. doi:10.2514/8.5282

[4] Stuart, D. (1962). The numerical solution of variational problems. *Journal of Mathematical Analysis and Applications*, 5, 30–45

[5] http://www.idsia.ch/~juergen/firstdeeplearner.html

[6] https://www.sciencedirect.com/science/article/pii/S0019995870904092/pdf?md5=e8000cc7c87531844fb9c884ff6054a5&pid=1-s2.0-S0019995870904092-main.pdf&_valck=1

[7] http://www.idsia.ch/~juergen/who-invented-backpropagation.html

[8] Ivakhnenko, A.G. (1971). Polynomial theory of complex systems. *IEEE Transactions on Systems, Man, and Cybernetics*, SMC-1(4), 364–378. doi:10.1109/tsmc.1971.4308320

[9] Fukushima, K. & Miyake, S. (1982). Neocognitron: A self-organizing neural network model for a mechanism of visual pattern recognition. *Lecture Notes in Biomathematics*, 267–285. doi:10.1007/978-3-642-46,466-9_18

[10] Hopfield, J.J. (1982). Neural networks and physical systems with emergent collective computational abilities. *Proceedings of the National Academy of Sciences*, 79(8), 2554–2558. doi:10.1073/pnas.79.8.2554

[11] Werbos, P.J. 1982. Applications of advances in nonlinear sensitivity analysis. *Lecture Notes in Control and Information Sciences*, 762–770. doi:10.1007/bfb0006203

[12] Ackley, D., Hinton, G. & Sejnowski, T. (1985). A learning algorithm for Boltzmann machines. *Cognitive Science*, 9(1), 147–169. doi:10.1016/s0364-0213(85)80012-4

[13] Rumelhart, D.E., Hinton, G.E. & Williams, R.J. (1986). Learning representations by back-propagating errors. *Nature*, 323(6088), 533–536. doi:10.1038/323533a0

[14] Smolensky, P. 1986. *Information processing in dynamical systems: Foundations of harmony theory* (CU-CS-321-86). Colorado University at the Boulder Department of Computer Science. February.

[15] LeCun, Y., Boser, B., Denker, J.S., Henderson, D., Howard, R.E., Hubbard, W. & Jackel, L.D. (1989). Backpropagation applied to handwritten zip code recognition. *Neural Computation*, 1(4), 541–551. doi:10.1162/neco.1989.1.4.541

[16] http://www.idsia.ch/~juergen/fundamentaldeeplearningproblem.html

[17] Hochreiter, S. & Schmidhuber, J. (1997). Long short-term memory. *Neural Computation*, 9(8), 1735–1780. doi:10.1162/neco.1997.9.8.1735

[18] Coates, A., Baumstarck, P., Le, Q. & Ng, A.Y. (2009). *Scalable learning for object detection with GPU hardware*, in *2009 IEEE/RSJ International Conference on Intelligent Robots and Systems*. doi:10.1109/iros.2009.5354084

[19] Deng, J., Dong, W., Socher, R., Li, L.-J., Kai, Li & Li, F.-F. (2009). *ImageNet: A large-scale hierarchical image database CPVR*, 13.

[20] Xavier, G., Antoine, B. & Yoshua, B. (2011). *Deep sparse rectifier neural networks*, in *Appearing in Proceedings of the 14th International Conference on Artificial Intelligence and Statistics (AISTATS) 2011*, Fort Lauderdale, FL, USA. Volume 15 of JMLR: W&CP 15. Copyright 2011 by the authors.

[21] Hinton, G.E., Srivastava, N., Krizhevsky, A., Sutskever, I. & Salakhutdinov, R.R. (2012). Improving neural networks by preventing co-adaptation of feature detectors. arXiv preprint arXiv:1207.0580.

[22] Ian, G., Jean, P.-A., Mehdi, M., Bing, X., David, W.-F., Sherjil, O., Aaron, C & Yoshua, B. (2014). Generative Adversarial Nets. Advances in Neural Information Processing Systems 27 (NIPS 2014)

[23] Silver, D., Huang, A., Maddison, C.J., Guez, A., Sifre, L., van den Driessche, G., … Hassabis, D. (2016). Mastering the game of Go with deep neural networks and tree search. *Nature*, 529(7587), 484–489. doi:10.1038/nature16961

[24] LeCun, Y., (2019). *1.1 Deep learning hardware: past, present, and future*, in *2019 IEEE International Solid-State Circuits Conference - (ISSCC)*. doi:10.1109/isscc.2019.8662396

[25] Hinton, G.E., Osindero, S. & Teh, Y.-W. (2006). A fast learning algorithm for deep belief nets. *Neural Computation*, 18(7), 1527–1554. doi:10.1162/neco.2006.18.7.1527

[26] https://kvssetty.com/fashionmnist-tutorial/

[27] Revathi, A.R., Rajalakshmi, P. & Shwettha, M. (2020).A neural network of human beings. ISSN 0970-647X. 44(8), November.

[28] https://towardsdatascience.com/building-a-convolutional-neural-network-male-vs-female-50347e2fa88b

[29] http://cstwiki.wtb.tue.nl/index.php?title=File:RNN-unrolled.png

[30] https://commons.wikimedia.org/wiki/File:Long_Short-Term_Memory.svg

[31] https://laptrinhx.com/restricted-boltzmann-machine-how-to-create-a-recommendation-system-for-movie-review-988466013/

[32] https://sigmoidal.io/beginners-review-of-gan-architectures/

[33] Sønderby, S.K. & Winther, O. Protein secondary structure prediction with long short term memory networks. arXiv1412.7828

[34] Rose, D.C., Arel, I., Karnowski, T.P. & Paquit, V.C. 2010. *"Applying deeplayered clustering to mammography image analytics*, in *Proceedings of the Biomedical Science Engineering Conferences.*, 1–4.

[35] Wang, J., MacKenzie, J.D., Ramachandran, R. & Chen, D.Z. 2016. *A deep learning approach for semantic segmentation in histology tissue images*, in *Proceedings of the MICCAI*, 176–184. [Online]. Available: http://dx.doi.org/10.1007/978-3-319-46723-8_21

[36] Lerouge, J., Herault, R., Chatelain, C., Jardin, F. & Modzelewski, R. 2015. Ioda: An input/output deep architecture for image labeling. *Pattern Recognition*, 48(9), 2847–2858.

[37] Zhou, Y. & Wei, Y. Jul. 2016. Learning hierarchical spectral-spatial features for hyperspectral image classification *IEEE Trans. Cybern.*, 46(7), 1667–1678.

[38] Pereira, S., Pinto, A., Alves, V. & Silva, C.A. 2016. Brain tumor segmentation using convolutional neural networks in MRI images. *IEEE Transactions on Medical Imaging*, 35, 1240–1251.

[39] Havaei, M., Davy, A., Warde-Farley, D., Biard, A., Courville, A. & Bengio, Y., et al. 2017. Brain tumor segmentation with deep neural networks. *Medical Image Analysis*, 35, 18–31.

[40] Shanand, J. & Li, L. 2016. *"A deep learning method for microaneurysm detection in fundus images"*, in *Proceedings of the IEEE Connected Health, Appl., Syst. Eng. Technol*, 357–358.

[41] Mansoor, A. et al. Aug. 2016. "Deep learning guided partitioned shape model for anterior visual pathway segmentation" *IEEE Transactions on Medical Imaging*, 35(8), 1856–1865.

[42] Li, F., Tran, L., Thung, K.H., Ji, S., Shen, D. & Li, J. Sep. 2015. A robust deep model for improved classification of ad/mci patients. *IEEE Journal of Biomedical and Health Informatics*, 19(5), 1610–1616.

[43] Wang, A., Song, C., Xu, X., Lin, F., Jin, Z. & Xu, W. Oct. 2015. *Selective and compressive sensing for energy-efficient implantable neural decoding*, in *Proc. IEEE Biomed. Circ. Syst. Conference*, 1–4.

[44] Wulsin, D., Blanco, J., Mani, R. & Litt, B. Dec. 2010. *Semi-supervised anomaly detection for eeg waveforms using deep belief nets*, in *Proceedings of the. 9th Int. Conf. Mach. Learn. Appl*, 436–441.

[45] Mirowski, P.W., Madhavan, D. & Lecun, Y. 2007. *Time-delay neural networks and independent component analysis for Eeg-Based prediction of epileptic seizures propagation*, in *Proc Conf AAAI Artif Intell*, 1892–1893.

[46] Mirowski, P., Madhavan, D., LeCun, Y. & Kuzniecky, R. 2009. Classification of patterns of EEG synchronization for seizure prediction. *Clinical Neurophysiology*, 120, 1927–1940.

[47] Petrosian, A., Prokhorov, D., Homan, R., Dasheiff, R. & Wunsch, D. 2000. Recurrent neural network based prediction of epileptic seizures in intra- and extracranial EEG. *Neurocomputing*, 30, 201–218.

[48] Xu, T., Zhang, H., Huang, X., Zhang, S. & Metaxas, D.N. 2016. *Multimodal deep learning for cervical dysplasia diagnosis,"* in *Proc. MICCAI*, 115–123. [Online]. Available: http://dx.doi.org/10.1007/978-3-31946723-8_14

[49] Avendi, M., Kheradvar, A. & Jafarkhani, H. 2016. A combined deep-learning and deformable-model approach to fully automatic segmentation of the left ventricle in cardiac mri. *Medical Image Analysis*, 30, 108–119.

[50] Yu, J.-s., Chen, J., Xiang, Z.,& Zou, Y.-X. 2015. *A hybrid convolutional neural networks with extreme learning machine for WCE image classification*, in *Proc. IEEE Robot. Biomimetics*, 1822–1827.

[51] Roth, H.R. et al. 2015. *Anatomy-specific classification of medical images using deep convolutional nets*, in *Proc. IEEE Int. Symp. Biomed. Imag*, 101–104.

[52] Van Grinsven, M.J., van Ginneken, B., Hoyng, C.B., Theelen, T. & S'anchez, C.I. May 2016. Fast convolutional neural network training using selective data sampling: application to hemorrhage detection in color fundus images. *IEEE Transactions on Medical Imaging*, 35(5), 1273–1284.

[53] Anthimopoulos, M., Christodoulidis, S., Ebner, L., Christe, A. & Mougiakakou, S. May 2016. Lung pattern classification for interstitial lung diseases using a deep convolutional neural network. *IEEE Transactions on Medical Imaging*, 35(5), 1207–1216.

[54] Roth, H.R., Farag, A., Lu, L., Turkbey, E.B. & Summers, R.M.. 2015. Deep convolutional networks for pancreas segmentation in CT imaging. *Med. Imag.: Image Proc.* (Vol. 9413, p. 94131G). International Society for Optics and Photonics.

[55] Prasoon, A., Petersen, K., Igel, C., Lauze, F., Dam, E. & Nielsen, M. 2013. Deep feature learning for knee cartilage segmentation using a triplanar convolutional neural network. *Med. Image Comput. Comput. Assist. Interv.*, 8150, 246–253.

[56] Kim, M., Wu, G. &Shen, D. 2013. Unsupervised deep learning for hippocampus segmentation in 7.0 T MR images. In: Wu G, Zhang D, Shen D, Yan P, Suzuki K & Wang F, editors. *Proceedings of the 4th International Workshop on Machine Learning in Medical Imaging*. New York: Springer-Verlag New York Inc.. 1–8.

[57] Ngo, T.A., Lu, Z. & Carneiro, G. 2017. Combining deep learning and level set for the automated segmentation of the left ventricle of the heart from cardiac cine magnetic resonance. *Medical Image Analysis*, 35, 159–171.

[58] Liao, S., Gao, Y., Oto, A. & Shen, D. 2013. Representation learning: a unified deep learning framework for automatic prostate MR segmentation. *Med Image Comput Comput Assist Interv*, 16, 254–261.

[59] Guo, Y.R., Wu, G.R., Commander, L.A., Szary, S., Jewells, V., Lin, W.L. et al. 2014. Segmenting hippocampus from infant brains by sparse patch matching with deep-learned features. *Med Image Comput Comput Assist Interv*, 8674, 308–315.

[60] Xu, Y., Zhu, J.Y., Chang, E. & Tu, Z. 2012. *Multiple clustered instance learning for histopathology cancer image classification, segmentation and clustering*, in *Proc IEEE Comput Soc Conf Comput Vis Pattern Recognit*, 964–971.

[61] Moeskops, P., Viergever, M.A., Mendrik, A.M., de Vries, L.S., Benders, M.J.N.L. & Isgum, I. 2016. Automatic segmentation of MR brain images with a convolutional neural network. *IEEE Transactions on Medical Imaging*, 35, 1252–1261.

[62] Nie, D., Zhang, H., Adeli, E., Liu, L. & Shen, D. 2016. *3d deep learning for multi-modal imaging-guided survival time prediction of brain tumor patients*. in *Proc. MICCAI*, 212–220. [Online]. Available: http://dx.doi.org/10.1007/978-3-319-46,723-8_25

[63] Kleesiek, J. et al. 2016. Deep MRI brain extraction: a 3D convolutional neural network for skull stripping. *NeuroImage*, 129, 460–469.

[64] Jiang, B., Wang, X., Luo, J., Zhang, X., Xiong, Y. & Pang, H. 2015. *Convolutional neural networks in automatic recognition of trans-differentiated neural progenitor cells under bright-field microscopy*, in *Proc. Instrum. Meas., Comput., Commun. Control*, 122–126.

[65] Havaei, M., Guizard, N., Larochelle, H. & Jodoin, P. 2016. *Deep learning trends for focal brain pathology segmentation in MRI*. In *Machine Learning for Health Informatics* (pp. 125–148). Springer, Cham.

[66] Suk, H.-I. et al. 2014. Hierarchical feature representation and multimodal fusion with deep learning for ad/mci diagnosis. *NeuroImage*, 101, 569–582.

[67] Kuang, D. & He, L. Nov. 2014. *Classification on ADHD with deep learning*, in *Proc. Cloud Comput. Big Data*, 27–32.

[68] Fritscher, K., Raudaschl, P., Zaffino, P., Spadea, M.F., Sharp, G.C. & Schubert, R., 2016, *Deep neural networks for fast segmentation of 3d medical images*, in *Proc. MICCAI*, 158–165. [Online]. Available: http://dx.doi.org/10.1007/978-3-319-46,723-8_19

[69] Zhen, X., Wang, Z., Islam, A., Bhaduri, M., Chan, I. & Li, S. 2016. Multi-scale deep networks and regression forests for direct bi-ventricular volume estimation. *Medical Image Analysis*, 30, 120–129.

[70] Brosch, T. et al. 2013. *Manifold learning of brain mris by deep learning,*" in *Proc. MICCAI*, 633–640.

[71] Ibrahim, R.,Yousri, N.A.,Ismail, M.A. & El-Makky, N.M. 2014. *"Multi-level gene/ mirna feature selection using deep belief nets and active learning,*" in *Proc. Eng. Med. Biol. Soc*, 3957–3960.

[72] Khademi, M. & Nedialkov, N.S. 2015. *Probabilistic graphical models and deep belief networks for prognosis of breast cancer,* in *Proc. IEEE 14th Int. Conf. Mach. Learn Appl,* 727–732.

[73] Quang, D., Chen, Y. & Xie, X. 2014. Dann: a deep learning approach for annotating the pathogenicity of genetic variants. *Bioinformatics,* 31, 761–763.

[74] Ramsundar, B., Kearnes, S., Riley, P., Webster, D., Konerding, D. & Pande, V. 2015. Massively multitask networks for drug discovery. ArXiv e-prints.

[75] Zhang, S. et al. 2016. A deep learning framework for modeling structural features of rna-bindissssssng protein targets. *Nucleic Acids Research,* 44(4), e32–e32.

[76] Tian, K., Shao, M., Zhou, S. & Guan, J. 2015. Boosting compound-protein interaction prediction by deep learning. *Methods,* 110, 29–34.

[77] Che, Z., Purushotham, S., Khemani, R. & Liu, Y. 2015. Distilling knowledge from deep networks with applications to healthcare domain, *ArXiv e-prints.*

[78] Miotto, R., Li, L., Kidd, B.A. & Dudley,J.T. 2016. Deep patient: an unsupervised representation to predict the future of patients from the electronic health records. *Scientific Reports,* 6, 1–10.

[79] Nie, L., Wang, M., Zhang, L.,Yan, S., Zhang, B. & Chua, T.S. 2015. Disease inference from health-related questions via sparse deep learning. *IEEE Transactions on Knowledge and Data Engineering,* 27(8), 2107–2119.

[80] Mehrabietal, S. 2015. *Temporal pattern and association discovery of diagnosis codes using deep learning,*" in *Proc. Int. Conf. Healthcare Informat,* 408–416.

[81] Shin, H., Lu, L., Kim, L., Seff, A., Yao, J. & Summers, R.M. 2015. Interleaved text/image deep mining on a large-scale radiology database for automated image interpretation. *CoRR,* vol. abs/1505.00670. [Online]. Available: http://arxiv.org/ abs/1505.00670

[82] Lipton,Z.C., Kale, D.C., Elkan, C. & Wetzel, R.C., 2015. Learning to diagnose with LSTM recurrent neural networks. *CoRR,* vol.abs/1511.03677. [Online]. Available: http://arxiv.org/abs/1511.03677

[83] Liang, Z., Zhang, G., Huang, J.X. & Hu, Q.V. 2014, *Deep learning for healthcare decision making with emrs,*" in *Proc. Int. Conf. Bioinformat. Biomed,* 556–559.

[84] Putin, E. et al. 2016. Deep biomarkers of human aging: application of deep neural networks to biomarker development. *Aging,* 8(5), 1–021.

[85] Futoma, J., Morris, J. & Lucas, J., 2015. A comparison of models for predicting early hospital readmissions. *Journal of Biomedical Informatics,* 56, 229–238.

[86] Wang, S., Weng, S., Ma, J. & Tang, Q. 2015. DeepCNF-D: predicting protein order/ disorder regions by weighted deep convolutional neural fields. *International Journal of Molecular Sciences,* 16, 17315–17330.

[87] Eickholt, J. & Cheng, J. 2013. DNdisorder: predicting protein disorder using boosting and deep networks. *BMC Bioinformatics,* 14, 88.

[88] Heffernan, R., Paliwal, K., Lyons, J., Dehzangi, A., Sharma, A., et al. 2015. Improving prediction of secondary structure, local backbone angles, and solvent accessible surface area of proteins by iterative deep learning. *Scientific Reports,* 5, 11476.

[89] Li, Y. & Shibuya ,T. 2015. *Malphite: a convolutieonal neural network and ensemble learning based protein secondary structure predictor. Proc. IEEE Int. Conf. Bioinformatics Biomed.,* 1260–1266.

[90] Lin, Z., Lanchantin, J. & Qi, Y. 2016. *MUST-CNN: a multilayer shift-and-stitch deep convolutional architecture for sequence-based protein structure prediction. Proc. Conf. AAAI Artif. Intell.*, 8.

[91] Troyanskaya, O.G. 2014. *Deep supervised and convolutional generative stochastic network for protein secondary structure prediction*, in *Proc. 31st Int. Conf. Mach. Learn*, 32, 745–753.

[92] Wang, S., Li, W., Liu, S. & Xu, J. 2016. RaptorX-Property: a web server for protein structure property prediction. *Nucleic Acids Research*, 44.

[93] Baldi, P., Pollastri, G., Andersen, C.A.F. & Brunak, S. 2000. *Matching protein beta-sheet partners by feedforward and recurrent neural networks*, in *Proc.Int. Conf. Intell. Syst. Mol. Biol* 25–36.

[94] Baldi, P., Brunak, S., Frasconi, P., Soda, G. & Pollastri, G. 1999. Exploiting the past and the future in protein secondary structure prediction. *Bioinformatics*, 15:937–946.

[95] Pollastri, G., Przybylski, D., Rost, B. & Baldi, P. 2002. Improving the prediction of protein secondary structure in three and eight classes using recurrent neural networks and profiles. *Proteins*, 47:228–235.

[96] Pollastri, G. & Baldi, P. 2002. Prediction of contact maps by GIOHMMs and recurrent neural networks using lateral propagation from all four cardinal corners. *Bioinformatics*, 18.

[97] Baldi, P. & Pollastri, G. 2004. The principled design of large-scale recursive neural network architectures-DAG-RNNs and the protein structure prediction problem. *Journal of Machine Learning Research*, 4:575–602.

[98] Di Lena P., Nagata K. & Baldi P. 2012. Deep architectures for protein contact map prediction. *Bioinformatics*, 28:2449–2457.

[99] Lyons, J., Dehzangi, A., Heffernan, R., Sharma, A., Paliwal, K., Sattar, A. et al. 2014. Predicting backbone Cα angles and dihedrals from protein sequences by stacked sparse auto-encoder deep neural network. *Journal of Computational Chemistry*, 35, 2040–2046.

[100] Spencer, M., Eickholt, J. & Cheng, J. 2015. A deep learning network approach to ab initio protein secondary structure prediction. *IEEE/ACM Transactions on Computational Biology and Bioinformatics*, 12, 103–112.

[101] Zhavoronkov, A. (2018). Artificial intelligence for drug discovery, biomarker development, and generation of novel chemistry. *Molecular Pharmaceutics*, 15(10), 4311–4313. doi:10.1021/acs.molpharmaceut.8b00930

[102] Eugene Bobrov, Anastasia Georgievskaya, Konstantin Kiselevl, Artem Sevastopolsky, Alex Zhavoronkov, Sergey Gurov, Konstantin Rudakov, Maria del Pilar Bonilla Tobar, Sören Jaspers & Sven Clemann. (2018). PhotoAgeClock: deep learning algorithms for development of non- invasive visual biomarkers of aging. *AGING 2018*, 10(11). www.aging-us.com

[103] Pepe, M.S., Etzioni, R., Feng, Z., Potter, J.D., Thompson, M.L., Thornquist, M., … Yasui, Y. (2001). phases of biomarker development for early detection of cancer. *JNCI Journal of the National Cancer Institute*, 93(14), 1054–1061. doi:10.1093/jnci/93.14.1054

[104] Ravi, D., et al. (2017). Deep learning for health informatics. *IEEE Journal of Biomedical and Health Informatics*, 21(1), 4–21.

[105] Shin et al. [27] proposed an image-text CNN to identify data that links images and reports of radiology from picture and information database of hospitals.

[106] Ong, B.T., Sugiura, K. & Zettsu, K. (2016). Dynamically pre-trained deep recurrent neural networks using environmental monitoring data for predicting PM2.5. *Neural Computing and Applications*, 27(6), 1553–1566.

[107] Zou, B., Lampos, V., Gorton, R. & Cox, I.J. (2016). On infect***ious intestinal disease surveillance using social media content, pp. 157–161

[108] Garimella, K., Alfayad, A. & Weber, I. (2015). *Social media image analysis for public health*, in: *Proceedings of the 2016 CHI Conference on Human Factors in Computing Systems*, 5543–5547.

[109] https://www.mayoclinic.org

[110] Stojan Trajanovski, Dimitrios Mavroeidis, Christine Leon Swisher, Binyam Gebrekidan Gebre, Bastiaan S. Veeling, Rafael Wiemker, Tobias Klinder, Amir Tahmasebi, Shawn M. Regis, Christoph Wald, Brady J. McKee, Sebastian Flacke, Heber MacMahon & Homer Pien (2019). Towards radiologist-level cancer risk assessment in CT lung screening using deep learning. arXiv:1804.01901v2 [cs.CV] 11 Apr 2019d

[111] Ardila, D., Kiraly, A.P., Bharadwaj, S., Choi, B., Reicher, J.J., Peng, L. & Shetty, S. (2019). End-to-end lung cancer screening with three-dimensional deep learning on low-dose chest computed tomography. *Nature Medicine*, 25(6), 954–961. doi:10.1038/s41591-019-0447-x

[112] Raja Subramanian, R., Nikhil Mourya, R., Prudhvi Teja Reddy, V., Narendra Reddy, B. & Srikar Amara (2020). Lung cancer prediction using deep learning framework. *International Journal of Control and Automation*, 13,(3), 154–160.

[113] Coudray, N., Ocampo, P. S., Sakellaropoulos, T., Narula, N., Snuderl, M., Fenyö, D. & Tsirigos, A. (2018). Classification and mutation prediction from non–small cell lung cancer histopathology images using deep learning. *Nature Medicine*. doi:10.1038/s41591-018-0177-5

[114] Xu, Y., Hosny, A., Zeleznik, R., Parmar, C., Coroller, T., Franco, I. & Aerts, H. J. W. L. (2019). Deep learning predicts lung cancer treatment response from serial medical imaging. *Clinical Cancer Research*. doi:10.1158/1078-0432.ccr-18-2495

[115] Mohamed Shakeel, P., Burhanuddin, M.A. & Desa, M.I. (2019). Lung cancer detection from CT image using improved profuse clustering and deep learning instantaneously trained neural networks. *Measurement*. doi:10.1016/j.measurement.2019.05.027

[116] Nasrullah Nasrullah, Jun Sang, Mohammad S. Alam, Muhammad Mateen, Bin Cai & Haibo Hu (2019). Automated lung nodule detection and classification using deep learning combined with multiple strategies. MDPI, *Sensors*, 19, 3722

[117] Lu, M.T., Raghu, V.K., Mayrhofer T., Aerts, H.J.W.L. & Udo Hoffmann (2020). Deep learning using chest radiographs to identify high-risk smokers for lung cancer screening computed tomography: development and validation of a prediction model. *Annals of Internal Medicine*, 173(9), 3 November 2020

[118] Yun Liu, Timo Kohlberger, Mohammad Norouzi, George E. Dahl, Jenny L. Smith, Arash Mohtashamian, Niels Olson, Lily H. Peng, Jason D. Hipp & Martin C. Stumpe. (2019). Artificial intelligence–Based breast cancer nodal metastasis detection. *Archives of Pathology & Laboratory Medicine*, 143, July 2019.

[119] Yala, A., Lehman, C., Schuster, T., Portnoi, T. & Barzilay, R. (2019). A deep learning mammography-based model for improved breast cancer risk prediction. *Radiology*, 292(1), 60–66. doi:10.1148/radiol.2019182716

[120] Shen, L., Margolies, L.R., Rothstein, J.H., Fluder, E., McBride, R. & Sieh, W. (2019). Deep learning to improve breast cancer detection on screening mammography. *Scientific Reports*, 9(1). doi:10.1038/s41598-019-48,995-4

[121] SanaUllah Khana, Naveed Islama, Zahoor Jana, Ikram Ud Dinb & Joel J.P.C. Rodrigues (2019). A novel deep learning based framework for the detection and classification of breast cancer using transfer learning. *Pattern Recognition Letters*, 125, 1 July 2019, 1–6.

[122] Xie, J., Liu, R., Luttrell, J. & Zhang, C. (2019). Deep learning based analysis of histo-pathological images of breast cancer. *Frontiers in Genetics*, 10. doi:10.3389/fgene.2019.00080

[123] Li-Qiang Zhou, Xing-Long Wu, Shu-Yan Huang, Ge-Ge Wu, Hua-Rong Ye, Qi Wei, Ling-Yun Bao, You-Bin Deng, Xing-Rui Li, Xin-Wu Cui & Christoph F. Dietrich (2019). Lymph node metastasis prediction from primary breast cancer us images using deep learning. *Radiology*, radiology.rsna.org.

[124] Hu, Q., Whitney, H.M. & Giger, M.L. (2020). A deep learning methodology for improved breast cancer diagnosis using multiparametric MRI. *Scientific Reports*, 10(1). doi:10.1038/s41598-020-67,441-4

[125] Bryan He, Ludvig Bergenstråhle, Linnea Stenbeck, Abubakar Abid, Alma Andersson, Åke Borg, Jonas Maaskola, Joakim Lundeberg & James Zou (2020). Integrating spatial gene expression and breast tumour morphology via deep learning. *Nature Biomedical Engineering*. www.nature.com/natbiomedeng.

[126] http://www.spatialtranscriptomicsresearch.org, https://wp.10xgenomics.com/spatial-transcriptomics

[127] https://www.healthline.com/health/heart-disease#types-of-heart-disease

[128] Medved, D., Ohlsson, M., Höglund, P., Andersson, B., Nugues, P. & Nilsson, J. (2018). Improving prediction of heart transplantation outcome using deep learning techniques. *Scientific Reports*, 8(1). doi:10.1038/s41598-018-21,417-7

[129] Poplin, R., Varadarajan, A.V., Blumer, K., Liu, Y., McConnell, M.V., Corrado, G.S. & Webster, D.R. (2018). Prediction of cardiovascular risk factors from retinal fundus photographs via deep learning. *Nature Biomedical Engineering*, 2(3), 158–164. doi:10.1038/s41551-018-0195-0

[130] Zreik, M., Lessmann, N. & van Hamersvelt, R.W., et al. (2018). Deep learning analysis of the myocardium in coronary CT angiography for identification of patients with functionally significant coronary artery stenosis. *Medical Image Analysis*, 44, 72–85.

[131] Ali, F., El-Sappagh, S., Islam, S. M. R., Kwak, D., Ali, A., Imran, M. & Kwak, K.-S. (2020). A smart healthcare monitoring system for heart disease prediction based on ensemble deep learning and feature fusion. *Information Fusion*. doi:10.1016/j.inffus.2020.06.008

[132] Shubhangi Khade, Anagha Subhedar, Kunal Choudhary, Tushar Deshpande & Unmesh Kulkarni (2019). A System to detect heart failure using deep learning techniques. *International Research Journal of Engineering and Technology (IRJET)*, 6(6), June 2019

[133] Kathleen H. Miao, & Julia H. Miao (2018). Coronary heart disease diagnosis using deep neural networks. *(IJACSA) International Journal of Advanced Computer Science and Applications*, 9(10).

[134] Mohsen, H., El-Dahshan, E.-S.A., El-Horbaty, E.-S.M. & Salem, A.-B.M. (2018). Classification using deep learning neural networks for brain tumors. *Future Computing and Informatics Journal*, 3(1), 68–71. doi:10.1016/j.fcij.2017.12.001

[135] Ali Ari & Davut Hanbay (2018). Deep learning based brain tumor classification and detection system. *Turkish Journal of Electrical Engineering and Computer Sciences*, 26, 2275–2286 © TÜBİTAK doi:10.3906/elk-1801-8

[136] Zahra Sobhaninia, Safiyeh Rezaei, Alireza Noroozi, Mehdi Ahmadi, Hamidreza Zarrabi, Nader Karimi, Ali Emami & Shadrokh Samavi (2018).Brain tumor segmentation using deep learning by type specific sorting of images. *Computer Vision and Pattern Recognition* (cs.CV); Image and Video Processing (eess.IV). arXiv:1809.07786

[137] Khan, M.A., Ashraf, I., Alhaisoni, M., Damaševičius, R., Scherer, R., Rehman, A. & Bukhari, S.A.C. (2020). Multimodal brain tumdsor classification using deep learning and robust feature selection: a machine learning application for radiologists. *Diagnostics*, 10(8), 565. doi:10.3390/diagnostics10080565

[138] Saba, T., Sameh Mohamed, A., El-Affendi, M., Amin, J. & Sharif, M. (2019). Brain tumor detection using fusion of hand crafted and deep learning features. *Cognitive Systems Research*. doi:10.1016/j.cogsys.2019.09.007

[139] https://www.nia.nih.gov/health/parkinsons-disease

[140] Camps, J., Samà, A., Martín, M., Rodríguez-Martín, D., Pérez-López, C., Moreno Arostegui, J.M. & Rodríguez-Molinero, A. (2018). Deep learning for freezing of gait detection in Parkinson's disease patients in their homes using a waist-worn inertial measurement unit. *Knowledge-Based Systems*, 139, 119–131. doi:10.1016/j.knosys.2017.10.017

[141] Rodríguez-Martín, D., Samà, A., Pérez-López, C., Català, A., Arostegui, J.M.M., Cabestany, J., Bayés, A., Alcaine, S., Mestre, B., Prats, A. et al. (2017) Home detection of freezing of gait using support vector machines through a single waist-worn triaxial accelerometer, *PLoS One*, 12(2), e0171764

[142] Wingate, J., Kollia, I., Bidaut, L. & Kollias, S., (2020). Unified deep learning approach for prediction of Parkinson's disease. *IET Image Processing*. doi:10.1049/iet-ipr.2019.1526

[143] Ramzi M. Sadek, Salah A. Mohammed, Abdul Rahman K. Abunbehan, Abdul Karim H. Abdul Ghattas, Majed R. Badawi, Mohamed N. Mortaja, Bassem S. Abu-Nasser & Samy S. Abu-Naser. (2019). Parkinson's Disease Prediction Using Artificial Neural Network. *International Journal of Academic Health and Medical Research (IJAHMR)* ISSN: 2000-007X 3(1), January.

[144] Tracy, J.M., Özkanca, Y., Atkins, D.C. & Ghomi, R.H.. (2019). Investigating voice as a biomarker: deep phenotyping methods for early detection of parkinson's disease. *Journal of Biomedical Informatics*, 103,362. doi:10.1016/j.jbi.2019.103362

[145] Wang, W., Lee, J., Harrou, F. & Sun, Y. (2020). Early detection of parkinson's disease using deep learning and machine learning. *IEEE Access*, 1–1. doi:10.1109/access.2020.3016062

[146] Van Steenkiste, T., Groenendaal, W., Deschrijver, D. & Dhaene, T. (2018). Automated sleep apnea detection in raw respiratory signals using long short-term memory neural networks. *IEEE Journal of Biomedical and Health Informatics*, 1. doi:10.1109/jbhi.2018.2886064

[147] Hafezi, M., Montazeri, N., Zhu, K., Alshaer, H., Yadollahi, A. & Taati, B. (2019). *Sleep apnea severity estimation from respiratory related movements using deep learning*, in *2019 41st Annual International Conference of the IEEE Engineering in Medicine and Biology Society (EMBC)*. doi:10.1109/embc.2019.8857524

[148] Feng, K., Qin, H., Wu, S., Pan, W. & Liu, G. (2020). A sleep apnea detection method based on unsupervised feature learning and single-lead electrocardiogram. *IEEE Transactions on Instrumentation and Measurement*, 1–1. doi:10.1109/tim.2020.3017246

[149] Li, K., Pan, W., Li, Y., Jiang, Q. & Liu, G. Jun. 2018. A method to detect sleep apnea based on deep neural network and hidden Markov model using single-lead ECG signal. *Neurocomputing*, 294, 94–101.

[150] Kale, D.C., Che, Z., Bahadori, M.T., Li, W., Liu, Y. & Wetzel, R. (2015). *Causal phenotype discovery via deep networks*, in *AMIA Annual Symposium Proceedings* https://www.ncbi.nlm.nih.gov/pmc/articles/PMC4765623/

2 Deep Knowledge Mining of Complete HIV Genome Sequences in Selected African Cohorts

Moses Effiong Ekpenyong, Mercy E. Edoho, Ifiok J. Udo, and Geoffery Joseph
University of Uyo, Nigeria

CONTENTS

DOI: 10.1201/9781003161233-2

2.1　INTRODUCTION

The human system is governed by unique and interesting organs that stir other bodily components. These organs (most of which are central in nature) are mainly symbiotic in that their functions and changes can be regarded as a sequence of organized chain-like reactions scripted by certain molecular components, most of which are largely genetic. Chief of these components is the DNA: Deoxyribonucleic acid–a molecule that holds genetic instructions used in the growth, development, functioning and reproduction of all living organisms, and indeed many viruses. The Human Immunodeficiency Virus (HIV) is a virus that spreads through certain body fluids and invades the immune system, especially the CD4 or T-cells. As the immune system degrades and becomes increasingly susceptible to several illnesses or opportunistic infections, the infection transforms into a more severe form known as the Acquired Immunodeficiency Syndrome (AIDS). Managing HIV/AIDS has posed numerous challenges such that available therapies are often not feasible in developing countries and economies due to high cost of treatment and incompetent health workers. Worse are the therapeutic constituents that form the required regimens–which may produce a myriad of clinical difficulties to efficient treatment. Should a regimen change be required due to resistance, ideally the new regimen formulation should contain at least two, or preferably three active drugs (Calvo and Daar 2014) including a careful review of the patient treatment history and resistance testing (to guide the choice of further medical treatment), (Riemenschneider and Heider 2016). In such cases, geno-type resistance testing becomes the right choice, as it is more cost effective than phenotype testing and can quickly detect resistance due to the presence of well-characterized mutations in the target viral gene(s).

There are basically 2 types of HIV: the type-1 HIV, also known as HIV-1, and the type-2 HIV, also called HIV-2. Whereas HIV-1 is the most common type of HIV and occurs all over the world, HIV-2 is mainly found in West Africa, but is slowly appearing in other regions, including the USA, Europe, and India. Although HIV-1 and HIV-2 are both retroviruses that present similar effects on the human body, they are however genetically distinct. In addition to human DNA sequences that directly interact with the integration machinery, the selection of HIV integration sites has also been shown to depend on the heterogeneous genomic context around a large region, which greatly hinders the prediction and mechanistic studies of HIV integration (Hu et al. 2019). Next-generation sequencing (NGS) has demonstrated good success for a variety of applications related to HIV patient monitoring, including the identification of drug resistance mutations (Fisher et al. 2015) and prediction of receptor tropism (Archer et al. 2012). However, full/complete genome sequences are also being exploited for detecting quasi-species and tracking immune evasion through variation in CD8 T lymphocyte epitopes with linkage elsewhere in the genome (Hughes et al. 2012).

Deep learning is fast becoming an effective method for highly accurate predictions from complex data sources, as convolutional neural networks continue to dominate classification problems. In the field of computational biology, deep learning appears a consistent prediction method in many applications, including the identification of nucleotide-protein binding and nucleotide modification sites (Alipanahi et al. 2015; Zhang et al. 2016; He et al. 2017; Li et al. 2017), prediction of the functional effects of non-coding sequence variants (Zhou and Troyanskaya 2015; Quang and Xie 2016), cancer genomics (Zhang et al. 2017a)), translation initiation and elongation modeling (Zhang et al. 2017b; Zhang et al. 2017c) and drug discovery (Wang and Zeng 2013; Wan and Zeng 2016). But despite the superior prediction performance of deep learning models, integrating the feature selection process and optimization of the network present some drawbacks, which not only limit its applicability but also raises potential concerns on the use of a 'black box'.

2.1.1 THE HIV-1 GENOME STRUCTURE

The complete sequence of the HIV-1 genome extracted from infectious virions has been resolved to a single-nucleotide resolution (Watts et al. 2009). The HIV genome encodes small number of viral proteins, unvaryingly establishing cooperative relations among HIV proteins and between HIV and the host proteins. This association is established to enable the invasion of host cells and the subsequent hijack of their internal machineries (Li G and De Clercq, 2016). HIV is called a retrovirus because it works in a reverse (back-to-front) manner. Unlike other viruses, retroviruses store their genetic information using RNA instead of DNA and need to 'make' DNA when they enter a human cell–to replicate or make new copies of themselves. In addition to viral replication, genetic and biologic variations of HIV are other properties that are linked directly to disease pathogenesis (Cheng-Mayer et al. 1988). Mutation in the reverse transcriptase (RT) gene selected by clinical treatment with azidothymidine (AZT), (Larder et al. 1989) or other RT inhibitors (Andries et al. 2004) demonstrates the biological importance of HIV-1 variation in vivo. Interestingly, the structure of HIV is different from other retroviruses (Temin 1989) with its molecular characterization revealing an exceptional level of sequence diversity (Berg et al. 2016). It is ~100nm in diameter, which innermost region consists of a cone-shaped core that includes two copies of the (positive sense) ssRNA genome, the enzymes reverse transcriptase, integrase and protease, some minor proteins, and the major core protein. The HIV genome encodes eight viral proteins playing essential roles during the HIV life cycle (Li and De Clercq 2016). Its type 1 variant (the HIV-1) contains two copies of non-covalently linked, un-spliced, positive-sense single-stranded RNA enclosed by a conical capsid composed of the viral protein p24, typical of lentiviruses (Montagnier 1999). The RNA component is 9,749 nucleotides long (Wain-Hobson et al. 1985) and bears a $5'$ cap (Gppp), a $3'$ poly (A) tail, and many open reading frames (ORFs), (Castelli and Levy 2002).

2.1.2 SPECIFIC OBJECTIVES AND CONTRIBUTION TO KNOWLEDGE

The specific objectives of this study include.

- To transform excavated HIV-1 genome sequences into digitally processed form, for efficient gene features extraction.
- To recognize inherent gene pattern expressions and establish natural clusters using unsupervised learning technique, for knowledge simplification and discovery.

- To mine cognitive knowledge from the established clusters, for supervised target output labeling.
- To learn the resultant dataset, for robust classification of new HIV-1 dataset.

This work has made novel contributions to knowledge, as follows:

Genome surveillance of new HIV cases: Molecular surveillance is crucial to monitor the genetic diversity of infectious diseases and for tracking emerging strains/sub-strains. This study presents a cognitive mining of existing HIV-1 features, for efficient cluster identification, and prompt decision support in the investigation of emerging HIV viral sub-strains.

Complete Genomic Sequence Processing: Most genome analysis methods use only one model to represent protein-coding regions in a genome, and hence are less likely to predict the location of genes that have an atypical sequence composition. This work enables the analysis of intra-genomic compositional variation in complete genome sequence. Using more than one machine learning technique, it is possible to process the entire genome datasets, for the detection of hidden patterns.

New approach to Viral Evolution Discovery: Viral evolution remains a main obstacle in the effectiveness of antiviral treatments. The ability to predict its evolution is important in the early detection of drug-resistant strains and potentially facilitates the design of more efficient antiviral treatments. This work offers an unsupervised neural computing approach to genome genealogy interpretation using pattern recognition. Our approach is certain to discover unexplored patterns or sub-strains not possible in trending bioinformatics tools.

Cooperative Approach to Genome Features Extraction and Prediction: Our approach is cooperative and provides a novel way of extracting gene features for genome analysis and prediction. It will offer useful support to gene-prediction software developers and genome annotation experts alike or as complement to existing gene prediction methods.

2.2   RELATED WORKS

2.2.1   CHARACTERIZATION OF HIV-1 SUBTYPES IN AFRICA

The apprehension of HIV-1 genetic variation and its effect on clinical management and drug resistance surveillance may proffer clinical relevance. Accordingly, Veras et al. (2011) used Bayesian phylodynamics to investigate the molecular epidemiology of CRF02_AG in Italy and African countries with emphasis on Cameroon. Among the 824 reverse transcript sequences retrieved from Los Alamos HIV databases, 70% of the 291 Cameroonian sequences clustered into three distinct clades (1, 2 and 3) while the remaining 30% intermixed with sequences from other African countries in the maximum likelihood tree. Clades 1 and 2 shared a common ancestor with sequences from West African, while Clade 3 identified with sequences from Gabon,

Ivory Coast, Mali, and Senegal. They finally evaluated the reliability of the tree using a Shimodaira–Hasegawa (SH)-like approximate likelihood ratio test (aLRT). Using the Phylip software package, Harris et al. (2003) phylogenetically analyzed 6 HIV-1 seropositive Ethiopian immigrants in Israel by neighborhood joining tree with reference genome from each subtype and two circulating recombinant forms (CRFs). The tree topography stability was evaluated by bootstrapping. They found that the six Ethiopian sequences clustered with strains of subtype C reference genome from Ethiopia and Israel. In Nigeria, Nazziwa et al. (2020) characterized 1,442 HIV-1 pol sequences from four geopolitical zones between 1999–2014 using maximum likelihood and Bayesian phylogenetic analysis. Subtypes were then determined by manual assignment with maximum likelihood phylogenetic analysis in Garli v0.98 using the general time reversible substitution model sequences. Their analysis revealed four prevalent strains: CRF02_AG (44%), CRF43_02G (16%), subtype G (8%) and CRF06_cpx (4%). Out of the 1,442 sequences, 328 sequences (23%) were unique recombinant forms (URFs). The 328 URFs majorly comprised of variants from three of the four prevalent subtypes circulating in Nigeria. The remaining percentages were minor variants of the virus. The study inferred a high prevalence of URFs and prospective CRFs. Khoja et al. (2008) analyzed 69 genome sequences collected from Aga Khan University Hospital in Nairobi, Kenya. The sequences were aligned to reference genome obtained from Los Alamos HIV database using Clustal X. The phylogenetic tree constructed using neighbor-joining method with PAUP revealed a predominant subtype A (39), subtype D (13), subtype C (7), subtype G (2). In addition to the subtypes, Simplot analysis identified recombinant types of AD (3), AC (1), AG (1) and CRF01_AE (3). Similarly, Billings et al. (2017) described the genetic diversity of HIV-1 among population in and around Mbeya, Tanzania. Multiple sequence alignment of HIV-1 isolates, and reference genome was generated using HIV Align but refined with MEGA version 5. Neighbor-joining phylogenetic trees were constructed, and bootstrap values computed using GTR + I + G model. Among the study isolates, subtypes C, A1 and D had the distribution 40%, 17% and 1% respectively while recombinants had the highest distribution of 42%. The incident strains compared to retrospective study showed no divergence from the earlier subtype distribution (43% C, 18% A1, 3% D and 36% recombinants). Investigation of HIV-1 subtype variability and phylogeny of 60 isolates from the Morocco national HIV sentinel surveillance program by Akrim et al. (2012), identified subtype B as the preponderant strain, out of which, 23% of the isolates comprised of non-B subtypes with CRF02_AG as the major CRF. A set of full reference genome sequences downloaded from Los Alamos database were aligned by codon with the study isolates using CLUSTAL algorithm. Analysis of the maximum likelihood tree was performed using the GTR + Gamma model of nucleotide evolution. Unlike previous studies that sequences only pol genes, Kemal et al. (2013) sequenced the gag, pol and env genes to investigate the subtype and inter-subtypes of HIV-1. The sequences were aligned with the reference strain from Los Alamos using BioEdit sequence alignment and analysis software. Individual phylogenetic tree was constructed for each of the three genes. HIV Subtypes and recombinant were determined using National Centre for Biotechnology Information (NCBI) and REGA BIOAFRICA tool version 2. Among the 30 isolates, 24 subtype A was found, four recombinants (AC and AD), one for subtype C and D each. In a prevailing subtype C strain in endemic South Africa, non-subtype C has mainly been reported in cosmopolitan cities of the country. In comparison to South African sequences downloaded from GenBank, Musyoki et al. (2015) identified and characterized a HIV-1 A1/C recombinant sub-strain in a female patient. The study involved the defragmentation of the genome sequences into individual genes (gag, vif, vpr, tat and rev, vpu) with the aid of sequence locator tool on the online HIV sequence database. Using REGA, jumping profile Hidden Markov Model (jpHMM), and neighbor-joining phylogenetic tree in MEGA 5 software package analytical tools, gag, vif, vpr, tat, and rev were assigned subtype A1 while vpu revealed subtype C. These diversities in subtypes and recombinant subtypes emphasized the need for continual surveillance of HIV genetics.

2.2.2 APPLICATION OF MACHINE LEARNING TO HIV FEATURE EXTRACTION AND PREDICTION

In Young et al. (2017), existing social media dataset associated with HIV and coded by an HIV domain expert, was tested on four commonly used machine learning models, to learn patterns associated with HIV risk behavior. They used 10-fold cross validation method to examine the speed and accuracy of these models in applying derived knowledge to detect HIV content in social media data. They found that Logistic regression and random forest gave highest accuracy and faster processing time. Kramer et al. (2001) were interested in molecular substructures or features that are frequent in the active molecules, and infrequent in the inactives. They mined features with minimum support in active compounds and a maximum support in inactive compounds. They analyzed the Developmental Therapeutics Program's AIDS antiviral screen database using the level-wise version space algorithm for inductive query. Yang and Chou (2003) mined biological data using a Self-Organizing Map (SOM). By partitioning a set of protein sequences, conventional homology alignment was applied to each cluster to determine the conserved local motif (or biological pattern) for the cluster. These local motifs were then regarded as rules for prediction and classification. Applying their approach to the prediction of HIV protease cleavage sites in proteins, they found that the rules derived from this method were much more robust than those derived from decision trees. Mahony et al. (2004) used a SOM to automatically identify multiple gene models within a genome. Their proposed implementation, called RescueNet, adopted relative synonymous codon as the indicator of protein-coding potential. Their implementation identified genes not detected by other methods. Cole and King (2013) developed a method to predict the disease prognosis of human HIV infections based on non-redundant HIV genomic and proteomic sequence data. Using the random forest on genomic data, they obtained over 91% accuracy at four different disease levels. They also analyzed the proteins expressed from five of the nine genes in HIV and found that the rev gene had the highest predictive performance for disease level. Using a decision tree, they were able to output rules that contained specific variants in the protein that can suggest disease outcomes. Singh, Narsai and Mars (2013) applied machine learning in the prediction of CD4 cell count of HIV-positive patient using genome sequences, viral load, and time. To accomplish this, a classification model was developed using support vector machine and neural network. Their results gave correlation co-efficient of 0.9 and classification accuracy of 95%.

Integrating HIV-1 genome to human genome is a crucial step in viral infection and replication cycle. Sükösd et al. (2015) developed a global RNA secondary structure model for the HIV-1 genome, which integrates both comparative structure analysis and information from experimental data in a full-length prediction without distance constraints. Besides recovering known structural elements, several novel structural elements that are conserved in HIV-1 evolution were predicted. They found that the structure of the HIV-1 genome was highly variable in most regions, with a limited number of stable and conserved RNA secondary structures. Salama et al. (2016) used machine learning technique to predict the possible point mutations that appear on alignments of primary RNA sequence structure. Their technique predicts the genotype of each nucleotide in the RNA sequence and proves the dependence of RNA sequence changes on nucleotides transitions. Neural network techniques were then utilized to predict new strains, and a rough set theory-based algorithm introduced to extract point mutation patterns. The algorithm was finally applied on several aligned RNA isolates time-series species of the Newcastle virus with two different datasets from two sources used in validating the study. Their results showed good nucleotide prediction accuracy, and the mutation rules were visualized for nucleotides correlation analysis in same RNA sequence. Hu et al. (2019) developed an attention-based deep learning framework for simultaneous prediction of HIV integration sites. Extensive tests on a high-density HIV integration site dataset showed the superiority of their frmework over conventional modeling strategies, through the automatic learning of genomic context of HIV integration solely from primary DNA sequence information. Systematic analysis on diverse known factors of HIV integration further validated the biological relevance of the predicted result. Steiner, Gibson, and Crandall (2020) utilized publicly available HIV-1 sequence data and drug resistance assay results for 18 antiretroviral therapy (ART) drugs to evaluate the performance of three architectures (multilayer perceptron, bidirectional recurrent neural network, and convolutional neural network) for drug resistance prediction, including biological analysis. They found that convolutional neural networks were the best performing architecture. Their results suggest that the high classification performance of deep learning models depend on drug resistance mutations (DRMs). Abe et al. (2020) developed an alignment-free clustering method based on an unsupervised neural network batch learning SOM where sequence fragments are clustered strictly based on oligonucleotide similarity without taxonomical information, to detect horizontal gene transfer (HGT) candidates and their origin in the entire genomes. Analysis of amino acid frequency suggested that housekeeping genes and some HGT candidates of the Antarctic strains exhibited different characteristics to other continental strains. Whereas Lys (Lysine), Ser (Serine), Thr (Threonine), and Val (Valine) increased in the Antarctic strains; Ala (Alanine), Arg (Arginine), Glu (Glutamic acid), and Leu (Leucine) showed noticeable decrease.

2.3 METHODS

2.3.1 DATA COLLECTION

Complete, high coverage sequences of over 8,500 bps were downloaded in the FASTA file format (a text-based format for representing either nucleotide sequences or amino acid or protein sequences, represented using single-letter codes) from the NCBI database (Fu et al. 2009) for patients infected with HIV-1. The present study is limited African countries, namely: East Africa (Ethiopia, Tanzania, Zambia, Malawi, Uganda, Kenya, Rwanda); West Africa (Nigeria), Central Africa (Cameroon), and South Africa (South Africa). A total of 30 genome isolates/sequences were selected from each country, totaling 300 genome sequences. Ambiguous nucleotide codes (K–Keto, G or T; M–Amino, A or C; N–Any nucleotide; R–Purine, A or G; S–Strong, G or C; V–Not T, A or G or C; W–Weak, A or T; and Y–Pyrimidine, C or T) besides A–Adenine, T–Thymine, C–Cytosine, and G–Guanine, present in the reference genome sequence, were ignored, such that the nucleotide positions are not shifted, as the presence of ambiguous nucleotides may potentially mask the genomic signature encoded within nucleotide frequencies. Total genome sequence length of (300 × 8,526 – 300 × 9,854) bps = (2,557,800 – 2,956,200) bps were excavated, processed, and stored in comma separated value (CSV) file.

2.3.2 PATTERN CLASSIFICATION ALGORITHM

The SOM is deployed in this study for learning the genome patterns. A SOM is an unsupervised neural network that implements nonlinear projection of multi-dimensional data onto a two-dimensional array of weight vectors. The purpose is to effectively preserve the topology of the high-dimensional data space presented by real-world problems. We repurpose the conventional SOM algorithm (Kiang 2001) for genome informatics and extend it for pattern clustering analysis using batch-learning SOM. Furthermore, we condition the learning process and resulting map to be independent of the order of data input. The initial weight vectors are random values, set and updated accordingly. Among various parameters associated with the SOM training, the neighborhood kernel is the most important one because it dictates the final topology of the trained map.

Let $\bar{x}$ be a gene expression vector defined by the expression levels of the gene to be analyzed. Each input vector is compared iteratively with the weights of the j-th output neuron, $\overline{w_j}$, searching for the neuron with the highest similarity with the gene behavior along the time course, organizing all the genes on the map according to their degree of similarity. The SOM algorithm is then performed in three main processes namely, competitive, cooperative, and adaptative.

The competitive process is carried out using competitive learning, such that only one output neuron wins (i.e., the vector with the maximum similarity to the input vector). The winning vector is called the best matching unit (BMU). An input vector contains the expression levels for all time samples for one gene. Maximum similarity can be computed with different functions, such as the Euclidean distance, the inner product, or the Mahalanobis distance functions, among others. The Euclidian distance function is used in this work, and is defined by,

$$\bar{E} = \left\| \bar{x} - \overrightarrow{w_j} \right\|$$ (2.1)

The input vector corresponding to the nucleotide transitions (expression levels) of one gene is compared against all the neurons in the output map to find the neuron with the maximum similarity. The BMU weight is adjusted as described below to obtain even more similarity with the input vector. In this manner, the possibility of being the winning vector during the competitive process on the next iteration is increased to be the neuron that represents the compared input vector on the output map.

The cooperative process computes which of the non-winning units are within the BMU's neighborhood. The weights of these units are also adjusted, but in proportion to their proximity to the BMU, while the weight of the units outside the neighborhood is left unchanged. To find the neighbor units, an initial neighborhood radius is set to monotonically shrink throughout the iterations. Useful neighborhood functions for neighborhood radius computation include the Mexican hat, the rectangular, and the Gaussian functions. In this work we deploy the Gaussian function, as it is the most widely used neighborhood function in SOMs, and is given by,

$$\tau_{ji} = \exp\left(\frac{d_i^2}{2\sigma^2} \right)$$ (2.2)

where τ_{ji} represents the topological neighborhood value of the unit j positioned around the winning unit; i, is the lateral distance between the winning unit i and the neighbor unit j; and σ is the decreasing neighborhood radius value.

A process of adaptation is finally imposed to maintain a stable topology and keep the neural network from overfitting. To achieve this, a learning rate based on the Hebbian hypothesis (η), which maintains a decreasing magnitude throughout the iterations is introduced. Every unit on the output map then adjusts its weight using the function,

$$\overrightarrow{w_j}(k+1) = \overrightarrow{w_j}(k) + \eta(k)\tau_{ji}(n)\left(\bar{x} - \overrightarrow{w_j} \right)$$ (2.3)

where k, is the number of iterations. The greater the proximity of neuron j to the winning neuron i, the higher the value of the neighborhood function $\tau_{ji}(k)$. This yields better adjustment on the weight of the neuron, as opposed to distanced ones farther away from the winning neuron. When the neuron is the BMU ($j = i$), the neighborhood function $\tau_{ji}(k)$ becomes unity, and the difference between the input vector and the weight of the BMU is multiplied by the learning rate $\eta(k)$ and added to its current weight. This increases the similarity of the BMU to the input vector. The desired behavior is the similarity between the neuron values on the output map and the input data.

2.3.3 GENOME FEATURE EXTRACTION

2.3.3.1 Nucleotide Transition Frequency

The HIV reference genome with accession number, AF033819.3 (the HIV-1 complete genome 9,181 bp linear RNA, obtained from the NCBI: www.ncbi.nlm.nih.gov) contains four conventional nucleotides, A, T, C, G, and as such, there are $4^2 = 16$ unique dinucleotide pairs that can be constructed from them, namely:

$$\omega = \{AA, AC, AG, AT, CA, CC, CG, CT, GA, GC, GG, GT, TA, TC, TG, TT\} \quad (2.4)$$

If we denote the frequency of the ith dinucleotide as d_i, then, a genomic sequence with 16-dimensional feature vector in the form of Equation (2.5) results,

$$f_\omega = \{d_{AA}, d_{AC}, d_{AG}, ..., d_{TT}\} \quad (2.5)$$

The frequencies of the dinucleotides are obtained by accumulating each dinucleotide transition along the extracted genome sequences. We ignore ambiguous nucleotides absent in the reference genome. Now, suppose n is the total length of a genome. By allowing a single sliding iteration window there exists $n - 1$ bubble counts. Hence, the dinucleotide frequency of d_i can be obtained by counting all nucleotide transitions that correspond to i.

2.3.3.2 Nucleotide Mutation Frequency

A pairwise alignment of each nucleotide with the reference genome was achieved by computing the frequency of mutated nucleotides down the sequence line. Owing to varying sequence lengths of the different genome isolates, a cutoff at the last nucleotide of the genome isolate or the reference genome serves as the maximum pair for comparison. Suppose n represents the total length of a genome. By permitting a single sliding iteration window, a mutation may be any of the following pair:

$$m = \{AC, AG, AT, CA, CG, CT, GA, GC, GT, TA, TC, TG\} \quad (2.6)$$

If we denote the frequency of the ith nucleotide pair as p_i, then, genomic sequence pairs with 12-dimensional feature vector in the form of Equation (2.7) results,

$$f_m = \{p_{AC}, p_{AG}, p_{AT}, ..., p_{TG}\} \quad (2.7)$$

2.3.3.3 Cognitive Knowledge Mining

Knowledge mining has served huge benefits for quick learning from big data. We apply Natural Language Processing of the genome datasets to extract knowledge of similar strains of the virus. A simple iteration technique is imposed on the SOM isolates ($i = 1, 2, 3, ..., n$), where n is the maximum number of isolates, as follows: For each isolate pattern, compile similar patterns with the rest of the isolates (i.e., $i + 1$, $i + 2, ..., n$). Concatenate compiled isolate(s) into a list $(j_1, j_2, ..., j_m)$ where j is an element of the list. Dump the compiled list into $CogMap(k_i \in j_1, j_2, ..., j_m)$.

2.3.3.4   Classification Algorithm

A deep neural network (DNN) is a hierarchical model where each layer implements a linear transformation accompanied by a non-linearity to the preceding layer. Let $X \in \mathbb{R}^{N \times D}$ represent the HIV genome feature dataset, with each row of X being a D-dimensional data point. For the sake of simplicity, we assume that the datasets lie in $\mathbb{R}$; and N is the number of training exemplars. Let $W^K \in \mathbb{R}^{d_{k-1} \cdot d_k}$ be a linearly transformed matrix applied to the output layer $k - 1$, $X_{K-1} \in \mathbb{R}^{N \cdot d_{k-1}}$, to produce a d_k-dimensional term $X_{K-1} W^K \in \mathbb{R}^{N \cdot d_k}$, at layer k. Suppose, $\varnothing_K : \mathbb{R} \to \mathbb{R}$ is a non-linear activation function, e.g., a sigmoid: $\varnothing_K(x) = (1 + e^{-x})^{-1}$ or hyperbolic tangent: $\varnothing_K = \tanh(x)$ or gaussian: $\varnothing_K = e^{-x2}$ or a rectified linear unit (ReLU): $\varnothing_K(x) = \max\{0, x\}$, then, an activation function can be applied to each instance of $Y_{K-1} W^K$ to generate the kth layer of a neural network: $X_K = \varnothing_K(X_{K-1} W^K)$. The output X_K of the network therefore becomes:

$$\gamma(X, W^1, W^2, ..., W^K) = \varnothing_K \left(\varnothing_{K-1} \left(... \varnothing_2 \left(\varnothing_1 \left(XW^1 \right) W^2 \right) ... W^{K-1} \right) W^K \right) \qquad (2.8)$$

Notice that: γ is an $N \times C$ matrix, and $C = d_k$, is the output network dimension, which equates to the number of classes of a classification problem. Hence, we can view γ as a function map that defines the input data X with fixed weights W.

2.4   RESULTS

2.4.1   IMPLEMENTATION WORKFLOW

The proposed implementation workflow executes a cooperative framework that mines the excavated genome sequences for robust classification of HIV-1 genome datasets. The steps implementing this framework as graphically demonstrated in Figure 2.1 using practical sample input and output data snippets, include, (i) Transform genome datasets; (ii) Visualize gene patterns; (iii) Extract cognitive knowledge; (iv) Extract feature dataset; (v) Visualize response clusters; (vi) Construct target outputs; (vii) Perform deep learning and classify/predict new genome dataset.

2.4.2   GENOME DATASET TRANSFORMATION

The excavated genome datasets were recoded according to the conventional nucleotide sequences, ignoring ambiguous nucleotides. Numerical values were then assigned to these nucleotides as follows: $A = 1$, $C = 3$, $G = 3$ and $T = 4$, while all ambiguous nucleotides were shifted up to permit a continuous sequence. The transformed sequences were then ready for SOM learning. A copy of the raw nucleotide sequences with ambiguous sequences were retained for the computation of the dinucleotide transition and nucleotide mutation frequencies.

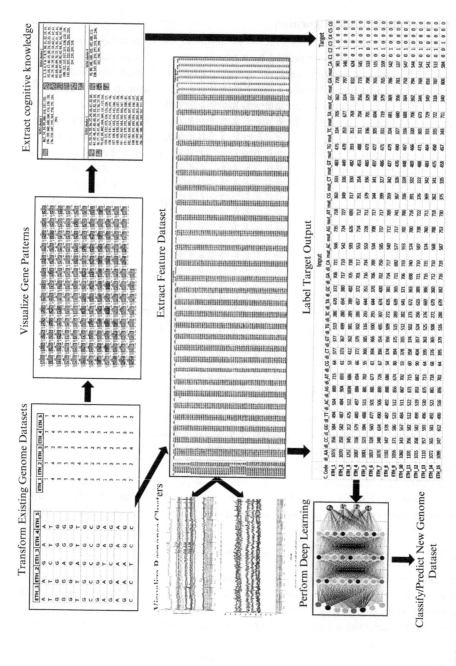

FIGURE 2.1 Proposed implementation workflow.

2.4.3   GENE PATTERN VISUALIZATION

During the SOM training, the map was linearly initialized along two greatest eigen-vectors of the input data. Then, the available nodes compete to win the input data, followed by updating the winner node and its topological neighbors. This iterative training was implemented using the batch algorithm and contains two phases: rough phase and fine-tuning phase. To increase the reproducibility of the trained map, we purposely prolonged the fine-turning phase until successive fine-tunings reached a steady state where the quality of the SOM map (i.e., average quantization error and topographic error) does not change any more. As regards parameters used in the SOM runs, the shape of the output map was defined as a sheet with a hexagonal lattice of 20 x 20 units, a Gaussian neighborhood function with an initial radius of three, and a decreasing learning rate type 'inv', were implemented in the MATLAB SOM Toolbox. The total number of iterations in the algorithm was defined as 10 multiplied by the number of neurons on the map divided by the number of genes to be analyzed. For effective visualization, the component planes of the SOM are presented in two separate displays (Figure 2.2a and b) with the reference genome occupying the first map space (see encircled).

Figure 2.2 explains the inherent pattern diversity in the HIV-1 genome sequences as well as between country (inter) and within country (intra) transmissions, as evidenced in the several patterns rendered by the component planes. This implies that some countries exhibit more than one pattern and show inter- and intra-similarities within and/or between countries.

2.4.4   COGNITIVE KNOWLEDGE EXTRACTION

To aid the cognitive extraction of dominant clusters and supervised labeling of target outputs, we disassembled the SOM correlation hunting matrix space and attribute these associations to different clusters. The outcome is Table 2.1–a cognitive link map with six clusters defining similarities in sub-strain categories or patterns.

2.4.5   FEATURE DATASET EXTRACTION

2.4.5.1   Response Cluster Visualization

Dinucleotide frequency for each genome sequence was computed for the 16 dinucleotide pairs within each genome sequence. The dinucleotide frequency feature vectors were then rendered as numerical values for all sequence analysis. A visualization of the response graph simulating the nucleotide transition frequency is given in Figure 2.3. Six (6) distinct clusters with dinucleotide transition frequencies ordered from high to low as follows: cluster 1 (AA), cluster 2 (AG), cluster 3 (GA, CA, AT, TA, GG), cluster 4 (AC, TG, TT), cluster 5 (CT, GC, GT, CC, TC) and cluster 6 (CG) were obtained.

Mutations continuously accumulate in HIV-1 genome after infection, rapidly evolving into a quasi-species population due to high level mutability of the virus and the host selection pressure. Mutation frequencies were computed for 12 nucleotide

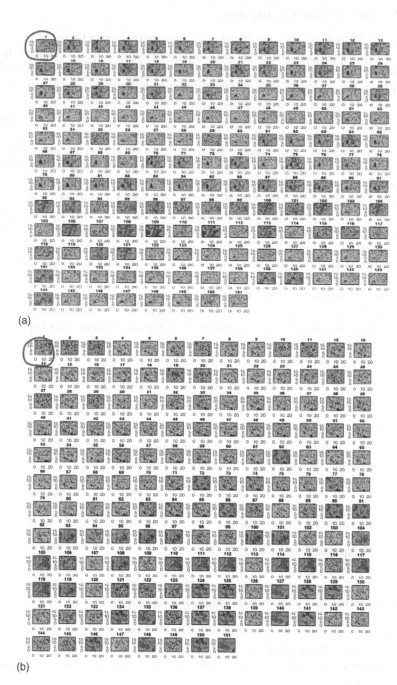

(a)

(b)

FIGURE 2.2 SOM component planes for excavated HIV-1 genome sequences.

TABLE 2.1
Cognitive Link Map of Isolate Clusters

SOM cluster 1 (Reference genome strain)

SOM cluster 2 (Sub-strain A)

SOM cluster 3 (Sub-strain B)

SOM cluster 4 (Sub-strain C)

SOM cluster 5 (Sub-strain D)

SOM cluster 6 (Sub-strain E)

pairs after a direct nucleotide alignment with the reference HIV-1 genome was achieved. The feature vectors were computed for all genome sequences used in this study and were rendered as numerical values for all sequence analysis. A visualization of the response graph simulating the nucleotide mutation frequency clusters is given in Figure 2.4. We notice variations in mutation among the genome sequences, and this depends on the sequence context (Geller et al. 2015). Furthermore, the response plots exhibit four distinct clusters with mutation frequencies ordered from

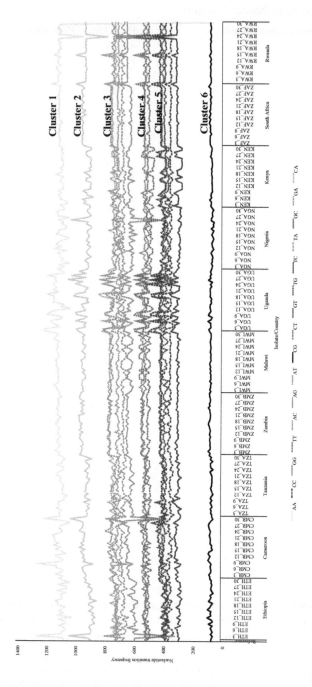

FIGURE 2.3   Clustering nucleotide transition frequency by country.

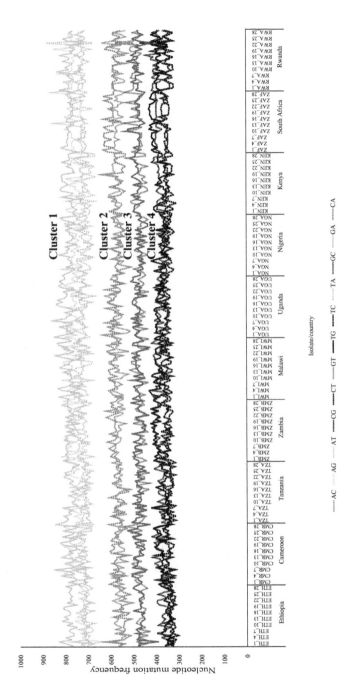

FIGURE 2.4 Clustering nucleotide mutation frequency by country.

high to low as follows: cluster 1 (GA, AG, TA, AT), cluster 2 (AC, CA), cluster C (GT, TG) and cluster 4 (CG, GC, CT, TC).

2.4.6 TARGET OUTPUT CONSTRUCTION

From the obtained clusters, target outputs were constructed using clusters discovered from the SOM analysis. Each cluster representing a viral sub-strain. Hence six clusters are supported in this study and were labeled accordingly as (C1, C2, C3, C4, C5 and C6). Each target class was coded as a binary (1 or 0) indicating the presence or absence of a sub-strain. These labels contribute to an enriched genome dataset with cognitive capabilities for informed supervised learning.

2.4.7 DNN CLASSIFICATION PERFORMANCE

We employed three common activation functions in a measure of our DNN model's performance. These functions include: the radial basis (radbas), sigmoid (logsig) and ReLU (poslin) activation functions. The sigmoid function is widely used because it is easily differentiable and facilitates backpropagation–as a lot of computation task is saved when training the network. Although sigmoid functions and their combinations work better for classification systems, they are always avoided due to vanishing and exploding gradients. While ReLU and max-pooling non-linearities are positively homogeneous, sigmoids are not, which explains why ReLU and max-pooling have shown improved performance depending on the solved problem. Besides, state-of-the-art networks are not trained with classical regularization such as ℓ_1 or ℓ_2 norm penalty on the weight parameters but rely more on dropout techniques (Srivastava et al. 2014).

MATLAB 2020a software was used to develop the necessary classification models for this study. Our DNN architecture (Figure 2.5) implements a pattern classifier with four hidden layers containing neurons in a dropout fashion as follows [24 18 12 6].

Confusion matrices were generated to enable a visualization of the proposed algorithm's performance. Figure 2.6a–c are overall confusion matrices implementing the gaussian, sigmoid and ReLU activation functions, respectively. In the confusion matrices, rows correspond to the predicted or output class, while columns correspond to the true or target class. The diagonal cells indicate correctly classified observations, and the off-diagonal cells correspond to incorrectly classified observations. The number of observations and percentage of the total number of observations are shown in each cell. The column on the far right shows the percentage of all samples predicted to belong to each class that are correctly and incorrectly classified–often

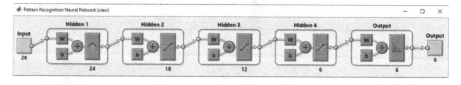

FIGURE 2.5   Proposed DNN architecture.

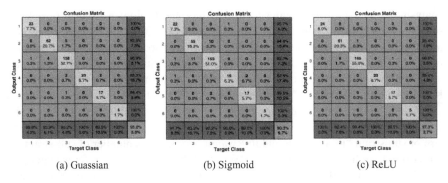

(a) Guassian   (b) Sigmoid   (c) ReLU

FIGURE 2.6   Confusion matrices of various activation functions.

TABLE 2.2
DNN Classification Performance for Various Activation Functions

Performance metric	Activation function		
	Gaussian	Sigmoid	ReLU
RMSE	0.0142	0.0278	0.0072
Classification Accuracy (%)	95.00	90.30	97.30
Precision	0.9454	0.9085	0.9605
Recall	0.9574	0.9194	0.9659
F1-Score	0.9514	0.9139	0.9632

called the precision, i.e.: TP/(TP + FP), where TP is the true positive and FP is the false positive. The row at the bottom of the plot shows the percentage of all samples belonging to each class that are correctly and incorrectly classified–often called the recall, i.e., TP/(TP + FN).

Table 2.2 shows the performance metrics obtained from learning the extracted genome features with various activation functions. The performance metrics used include the root mean squared error (RMSE), classification accuracy, precision, and recall. Aside the RMSE, other metrics were obtained from the confusion matrices. Observe that the ReLU activation function gave the best performance measure when compared with other activation functions.

2.4.8 FINDINGS AND SCIENTIFIC IMPLICATION OF THE STUDY

While the use of bioinformatic tools is widely accepted for plotting the phylogeny of viral strains at various location sites or subclade regions, very few attempts have been made to efficiently exploit machine learning techniques in the identification and clustering of complete genome sequences for sub-strain discovery. Available literature concentrates on learning population patterns associated with risk behaviors and the prediction of protease cleavage sites in proteins as well as disease surveillance. This work is therefore novel as we introduce a robust SOM algorithm that learns a transformed HIV-1 dataset including the reference genome for sub-strains (dis)similarity analysis. From the results obtained in this study, the Ethiopian isolates revealed 2 separate sub-strains or patterns "A" and "B" different from the reference genome, implying the evolution of more stable sub-strains and are likened to other Ethiopian isolates, save isolates 5, 15, 18, 19, 22, 24, 25 and 26. Sub-strain A also occur sparingly in Cameroon in two isolates, but are more prevalent in Tanzania, Zambia, and Kenya. A new sub-strain "C", typical in Rwandan isolates is also found in Zambia and Kenya, while another novel sub-strain "D" dominates mainly South Africa, with 2 premature copies creeping into Rwanda. A less dominant, but newly emerging pattern or sub-strain "E" is also found in Rwanda, bringing its number of sub-strains to four, excluding the reference genome. We found that copies of the reference genome are still being retained by isolates within Africa. Countries still retaining sequence of the reference genome strain include Cameroon (20 isolates), Zambia (3 isolates), Uganda (10 isolates), Kenya (2 isolates), South Africa (4 isolates) and Rwanda (3 isolates). Aside the reference genome, Uganda (20/30) and Nigeria (30/30) still maintain only one sub-strain of the virus. The scientific implication of this confirms the existence of inter- and intra- sub-strain transmission and requires more research progress in the efficient prediction of emerging sub-strains, which this research has opened.

Detection of GA (transitive) hypermutation in HIV-1 and HIV-2 was confirmed in Rawson (2015). Also, transversive mutations AG show same frequency, and were 10 times more frequent than CG. Our study validates this claim, as Figure 2.4 reveals frequency similarities of the GA/AG, AC/CA, and CT/TC mutants, and a 10-fold increase in nucleotide mutation frequency between mutant AG and CG. Strangely, the behavior of the CG and GC mutants are not in anywhere similar, as the CG mutant appears to experience very high mutations, indicating rapid gene replications compared to GC. The scientific implication of this exception may be due to the type of treatment as reported in Mullins et al. (2011).

Given the computation intensity of our problem, the ReLU activation function was found to outsmart the other activation functions, but more research in this direction is required to improve our classification system and make the system more appropriate for decision support systems.

2.5 CONCLUSION

Past contributions of deep learning approaches to HIV genome sequences study are limited to prediction of new virus strains, drug resistance risk and extraction of point mutation patterns among others. This study provided adequate knowledge that informs robust viral evolution analysis with a view of revealing the subtype of HIV genome retained by emerging isolates. It could also assist researchers in the efficient extraction of HIV genome features for detection of inherent patterns required for the formation of natural clusters. This study implemented a deep knowledge mining framework by combining a batch-learning self-organizing map (SOM) and DNN classifier. The suitability of SOM in learning genome patterns outweighs decision tree approach, by its ability to implement nonlinear projection of multi-dimensional data onto a 2-dimensional array of weight vectors, to effectively preserve the topology of the high-dimensional data space presented by real-world problems. To enhance the performance of our SOM, we modified the conventional SOM using 3 major processes, namely, competitive, cooperative, and adaptive. Out of over 8,500 bp, representing each nucleotide, amino acids or protein sequence, a total of 30 isolates or genome sequences with length ranging from 2,557,800 to 2,956,200 bps were selected from 10 African countries under study. Based on six target output labels which provided enriched genome dataset with cognitive capabilities for informed supervised learning, DNN classification was performed. Results of SOM and cognitive mining proves the existence of sub-strains different from the reference genomes in selected African countries and the prospects of improved genome surveillance for efficient contact tracing of infectious diseases. Classification results show the superiority of the ReLU activation function in learning genome features.

REFERENCES

Abe, Takashi, Yu Akazawa, Atsushi Toyoda, Hironori Niki, and Tomoya Baba. "Batch-learning self-organizing map identifies horizontal gene transfer candidates and their origins in entire genomes." *Frontiers in Microbiology* 11 (2020): 1486. https://doi.org/10.3389/fmicb.2020.01486.

Akrim, Mohammed, Sanae Lemrabet, Elmir Elharti, Rebecca R. Gray, Jean Claude Tardy, Robert L. Cook, Marco Salemi, Patrice Andre, Taj Azarian, and Rajae El Aouad. "HIV-1 Subtype distribution in morocco based on national sentinel surveillance data 2004–2005." *AIDS Research and Therapy* 9, no. 1 (2012): 1–8. https://doi.org/10.1186/1742-6405-9-5.

Alipanahi, Babak, Andrew Delong, Matthew T. Weirauch, and Brendan J. Frey. "Predicting the sequence specificities of DNA-and RNA-binding proteins by deep learning." *Nature Biotechnology* 33, no. 8 (2015): 831–838. https://doi.org/10.1038/nbt.3300.

Andries, Koen, Hilde Azijn, Theo Thielemans, Donald Ludovici, Michael Kukla, Jan Heeres, Paul Janssen et al. "TMC125, a novel next-generation nonnucleoside reverse transcriptase inhibitor active against nonnucleoside reverse transcriptase inhibitor-resistant human immunodeficiency virus type 1." *Antimicrobial Agents and Chemotherapy* 48, no. 12 (2004): 4680–4686. https://doi.org/10.1128/AAC.48.12.4680-4686.2004.

Archer, John, Jan Weber, Kenneth Henry, Dane Winner, Richard Gibson, Lawrence Lee, Ellen Paxinos et al. "Use of four next-generation sequencing platforms to determine HIV-1 coreceptor tropism." *PLoS One* 7, no. 11 (2012): e49602. https://doi.org/10.1371/journal.pone.0049602.

Berg, Michael G., Julie Yamaguchi, Elodie Alessandri-Gradt, Robert W. Tell, Jean-Christophe Plantier, and Catherine A. Brennan. "A pan-HIV strategy for complete genome sequencing." *Journal of Clinical Microbiology* 54, no. 4 (2016): 868–882. https://doi.org/10.1128/JCM.02479-15.

Billings, Erik, Eric Sanders-Buell, Meera Bose, Gustavo H. Kijak, Andrea Bradfield, Jacqueline Crossler, Miguel A. Arroyo et al. "HIV-1 genetic diversity among incident infections in Mbeya, Tanzania." *AIDS Research and Human Retroviruses* 33, no. 4 (2017): 373–381. https://doi.org/10.1089/aid.2016.0111.

Calvo, Katya R., and Eric S. Daar. "Antiretroviral therapy: Treatment-experienced individuals." *Infectious Disease Clinics* 28, no. 3 (2014): 439–456. https://doi.org/10.1016/j.idc.2014.06.005.

Castelli, J.C., and A. Levy (2002). HIV (Human Immunodeficiency Virus). *Encyclopedia of Cancer*. (2nd ed.). p. 407–415.

Cheng-Mayer, Cecilia, Deborah Seto, Masatoshi Tateno, and Jay A. Levy. "Biologic features of HIV-1 that correlate with virulence in the host." *Science* 240, no. 4848 (1988): 80–82. https://doi.org/10.1126/science.2832945.

Cole, Charles L., and Brian R. King. *"Using Machine Learning to Predict the Health of HIV-Infected Patients."* In *Proceedings of the International Conference on Bioinformatics, Computational Biology and Biomedical Informatics*, pp. 684–685. 2013. https://doi.org/10.1145/2506583.2506685.

Fisher, Randall G., Davey M. Smith, Ben Murrell, Ruhan Slabbert, Bronwyn M. Kirby, Clair Edson, Mark F. Cotton, Richard H. Haubrich, Sergei L. Kosakovsky Pond, and Gert U. Van Zyl. "Next generation sequencing improves detection of drug resistance mutations in infants after PMTCT failure." *Journal of Clinical Virology* 62 (2015): 48–53. https://doi.org/10.1016/j.jcv.2014.11.014.

Fu, William, Brigitte E. Sanders-Beer, Kenneth S. Katz, Donna R. Maglott, Kim D. Pruitt, and Roger G. Ptak. "Human immunodeficiency virus type 1, human protein interaction database at NCBI." *Nucleic Acids Research* 37, no. suppl_1 (2009): D417–D422. https://doi.org/10.1093/nar/gkn708.

Geller, Ron, Pilar Domingo-Calap, José M. Cuevas, Paola Rossolillo, Matteo Negroni, and Rafael Sanjuán. "The external domains of the HIV-1 envelope are a mutational cold spot." *Nature Communications* 6, no. 1 (2015): 1–9. https://doi.org/10.1038/ncomms9571.

Harris, Matthew E., Shlomo Maayan, Bohye Kim, Michael Zeira, Guido Ferrari, Deborah L. Birx, and Francine E. McCutchan. "A cluster of HIV type 1 subtype C sequences from Ethiopia, observed in full genome analysis, is not sustained in subgenomic regions." *AIDS Research and Human Retroviruses* 19, no. 12 (2003): 1125–1133. https://doi.org/10.1089/088922203771881220.

He, Xuan, Sai Zhang, Yanqing Zhang, Tao Jiang, and Jianyang Zeng. "Characterizing RNA pseudouridylation by convolutional neural networks." *bioRxiv* (2017): 126979. https://doi.org/10.1101/126979.

Hu, Hailin, An Xiao, Sai Zhang, Yangyang Li, Xuanling Shi, Tao Jiang, Linqi Zhang, Lei Zhang, and Jianyang Zeng. "DeepHINT: Understanding HIV-1 integration via deep learning with attention." *Bioinformatics* 35, no. 10 (2019): 1660–1667. https://doi.org/10.1093/bioinformatics/bty842.

Hughes, Austin L., Ericka A. Becker, Michael Lauck, Julie A. Karl, Andrew T. Braasch, David H. O'Connor, and Shelby L. O'Connor. "SIV genome-wide pyrosequencing provides a comprehensive and unbiased view of variation within and outside CD8 T lymphocyte epitopes." *PLoS One* 7, no. 10 (2012): e47818. https://doi.org/10.1371/journal.pone.0047818

Kemal, Kimdar S., Kathryn Anastos, Barbara Weiser, Christina M. Ramirez, Qiuhu Shi, and Harold Burger. "Molecular epidemiology of HIV type 1 subtypes in Rwanda." *AIDS Research and Human Retroviruses* 29, no. 6 (2013): 957–962. https://doi.org/10.1089/aid.2012.0095.

Khoja, Suhail, Peter Ojwang, Saeed Khan, Nancy Okinda, Reena Harania, and Syed Ali. "Genetic analysis of HIV-1 subtypes in Nairobi, Kenya." *PLoS One* 3, no. 9 (2008): e3191. https://doi.org/10.1371/journal.pone.0003191.

Kiang, Melody Y. "Extending the Kohonen self-organizing map networks for clustering analysis." *Computational Statistics & Data Analysis* 38, no. 2 (2001): 161–180. https://doi.org/10.1016/S0167-9473(01)00040-8.

Kramer, Stefan, Luc De Raedt, and Christoph Helma. *"Molecular feature mining in HIV data."* In *Proceedings of the seventh ACM SIGKDD international conference on Knowledge discovery and data mining*, pp. 136–143. 2001. https://doi.org/10.1145/502512.502533

Larder, Brendan A., Graham Darby, and Douglas D. Richman. "HIV with reduced sensitivity to zidovudine (AZT) isolated during prolonged therapy." *Science* 243, no. 4899 (1989): 1731–1734. https://doi.org/10.1126/science.2467383.

Li, Guangdi, and Erik De Clercq. "HIV genome-wide protein associations: A review of 30 years of research." *Microbiology and Molecular Biology Reviews* 80, no. 3 (2016): 679–731. https://doi.org/10.1128/MMBR.00065-15.

Li, Shuya, Fanghong Dong, Yuexin Wu, Sai Zhang, Chen Zhang, Xiao Liu, Tao Jiang, and Jianyang Zeng. "A deep boosting based approach for capturing the sequence binding preferences of RNA-binding proteins from high-throughput CLIP-seq data." *Nucleic Acids Research* 45, no. 14 (2017): e129-e129. https://doi.org/10.1093/nar/gkx492

Mahony, Shaun, James O. McInerney, Terry J. Smith, and Aaron Golden. "Gene prediction using the self-organizing map: Automatic generation of multiple gene models." *BMC Bioinformatics* 5, no. 1 (2004): 1–9. https://doi.org/10.1186/1471-2105-5-23.

Montagnier, Luc. "Human immunodeficiency viruses." In *Encyclopedia of Virology* (Webster, R. G. & Granoff, A., eds), (1999), pp. 763–774, Academic Press, SanDiego, USA.

Mullins, James I., Laura Heath, James P. Hughes, Jessica Kicha, Sheila Styrchak, Kim G. Wong, Ushnal Rao et al. "Mutation of HIV-1 genomes in a clinical population treated with the mutagenic nucleoside KP1461." *PLoS One* 6, no. 1 (2011): e15135. https://doi.org/10.1371/journal.pone.0015135.

Musyoki, Andrew M., Johnny N. Rakgole, Gloria Selabe, and Jeffrey Mphahlele. "Identification and genetic characterization of unique HIV-1 A1/C recombinant strain in South Africa." *AIDS Research and Human Retroviruses* 31, no. 3 (2015): 347–352. https://doi.org/10.1089/aid.2014.0212.

Nazziwa, Jamirah, Nuno Rodrigues Faria, Beth Chaplin, Holly Rawizza, Phyllis Kanki, Patrick Dakum, Alash'le Abimiku, Man Charurat, Nicaise Ndembi, and Joakim Esbjörnsson. "Characterisation of HIV-1 molecular epidemiology in nigeria: Origin, diversity, demography and geographic spread." *Scientific Reports* 10 (2020). https://doi.org/10.1038/s41598-020-59944-x.

Quang, Daniel, and Xiaohui Xie. "DanQ: A hybrid convolutional and recurrent deep neural network for quantifying the function of DNA sequences." *Nucleic Acids Research* 44, no. 11 (2016): e107–e107.https://doi.org/10.1093/nar/gkw226.

Rawson, Jonathan M. O., Sean R. Landman, Cavan S. Reilly, and Louis M. Mansky. "HIV-1 and HIV-2 exhibit similar mutation frequencies and spectra in the absence of G-to-A hypermutation." *Retrovirology* 12, no. 1 (2015): 1–17. https://doi.org/10.1186/s12977-015-0180-6

Riemenschneider, M., and D. Heider. "Current approaches in computational drug resistance prediction in HIV. *Current HIV Research* 14, no. 4 2016: 307–315.

Salama, Mostafa A., Aboul Ella Hassanien, and Ahmad Mostafa. "The prediction of virus mutation using neural networks and rough set techniques." *EURASIP Journal on Bioinformatics and Systems Biology* 2016, no. 1 (2016): 10. https://doi.org/10.1186/s13637-016-0042-0.

Singh, Yashik, Nitesh Narsai, and Maurice Mars. "Applying machine learning to predict patient-specific current CD 4 cell count in order to determine the progression of human immunodeficiency virus (HIV) infection." *African Journal of Biotechnology* 12, no. 23 (2013). https://doi.org/0.5897/AJB12.1860.

Srivastava, Nitish, Geoffrey Hinton, Alex Krizhevsky, Ilya Sutskever, and Ruslan Salakhutdinov. "Dropout: A simple way to prevent neural networks from overfitting." *The Journal of Machine Learning Research* 15, no. 1 (2014): 1929-1958.

Steiner, Margaret C., Keylie M. Gibson, and Keith A. Crandall. "Drug resistance prediction using deep learning techniques on HIV-1 sequence data." *Viruses* 12, no. 5 (2020): 560. https://doi.org/10.3390/v12050560.

Sükösd, Zsuzsanna, Ebbe S. Andersen, Stefan E. Seemann, Mads Krogh Jensen, Mathias Hansen, Jan Gorodkin, and Jørgen Kjems. "Full-length RNA structure prediction of the HIV-1 genome reveals a conserved core domain." *Nucleic Acids Research* 43, no. 21 (2015): 10168–10179. https://doi.org/10.1093/nar/gkv1039.

Temin, Howard M. "Is HIV unique or merely different?" *JAIDS Journal of Acquired Immune Deficiency Syndromes* 2, no. 1 (1989): 1–9.

Véras, Nazle Mendonca Collaço, Maria Mercedes Santoro, Rebecca R. Gray, Andrew J. Tatem, Alessandra Lo Presti, Flaminia Olearo, Giulia Cappelli et al. "Molecular epidemiology of HIV type 1 CRF02_AG in Cameroon and African patients living in Italy." *AIDS Research and Human Retroviruses* 27, no. 11 (2011): 1173–1182. https://doi.org/10.1089/aid.2010.0333.

Wain-Hobson, Simon, Pierre Sonigo, Olivier Danos, Stewart Cole, and Marc Alizon. "Nucleotide sequence of the AIDS virus, LAV." *Cell* 40, no. 1 (1985): 9–17. https://doi.org/doi:10.1016/0092-8674(85)90303-4.

Wan, Fangping, and Jianyang Michael Zeng. "Deep learning with feature embedding for compound-protein interaction prediction." *bioRxiv* (2016): 086033. https://doi.org/10.1101/086033.

Wang, Yuhao, and Jianyang Zeng. "Predicting drug-target interactions using restricted Boltzmann machines." *Bioinformatics* 29, no. 13 (2013): i126–i134. https://doi.org/10.1093/bioinformatics/btt234

Watts, Joseph M., Kristen K. Dang, Robert J. Gorelick, Christopher W. Leonard, Julian W. Bess Jr, Ronald Swanstrom, Christina L. Burch, and Kevin M. Weeks. "Architecture and secondary structure of an entire HIV-1 RNA genome." *Nature* 460, no. 7256 (2009): 711–716. https://doi.org/10.1038/nature08237.

Yang, Zheng Rong, and Kuo-Chen Chou. "Mining biological data using self-organizing map." *Journal of Chemical Information and Computer Sciences* 43, no. 6 (2003): 1748–1753. https://doi.org/10.1021/ci034138n.

Young, Sean D., Wenchao Yu, and Wei Wang. "Toward automating HIV identification: Machine learning for rapid identification of HIV-related social media data." *Journal of Acquired Immune Deficiency Syndromes* 74, no. Suppl 2 (2017): S128. https://doi.org/10.1097/QAI.0000000000001240.

Zhang, Sai, Hailin Hu, Jingtian Zhou, Xuan He, Tao Jiang, and Jianyang Zeng. "Analysis of ribosome stalling and translation elongation dynamics by deep learning." *Cell Systems* 5, no. 3 (2017b): 212–220. https://doi.org/10.1016/j.cels.2017.08.004.

Zhang, Sai, Hailin Hu, Tao Jiang, Lei Zhang, and Jianyang Zeng. "TITER: Predicting translation initiation sites by deep learning." *Bioinformatics* 33, no. 14 (2017c): i234–i242. https://doi.org/10.1093/bioinformatics/btx247.

Zhang, Sai, Jingtian Zhou, Hailin Hu, Haipeng Gong, Ligong Chen, Chao Cheng, and Jianyang Zeng. "A deep learning framework for modeling structural features of RNA-binding protein targets." *Nucleic Acids Research* 44, no. 4 (2016): e32–e32. https://doi.org/10.1093/nar/gkv1025.

Zhang, Sai, Muxuan Liang, Zhongjun Zhou, Chen Zhang, Ning Chen, Ting Chen, and Jianyang Zeng. "Elastic restricted Boltzmann machines for cancer data analysis." *Quantitative Biology* 5, no. 2 (2017a): 159–172. https://doi.org/10.1007/s40484-017-0092-7.

Zhou, Jian, and Olga G. Troyanskaya. "Predicting effects of noncoding variants with deep learning–based sequence model." *Nature Methods* 12, no. 10 (2015): 931–934. https://doi.org/10.1038/nmeth.3547.

3 Review of Machine Learning Approach for Drug Development Process

Devottam Gaurav
Indian Institute of Technology, India

Fernando Ortiz Rodriguez and Sanju Tiwari
Universidad Automata de Tamaulipas, Mexico

M.A. Jabbar
Vardhaman College of Engineering, India

CONTENTS

DOI: 10.1201/9781003161233-3

3.1 INTRODUCTION

Artificial intelligence (AI) is the recreation of the human knowledge measure by PCs. The cycle incorporates getting data, creating rules for utilizing the data, drawing surmised or positive ends, and self-amendment. The progression of AI can be viewed as a twofold: many dread that it will compromise their business; paradoxically, every development in AI is praised in view of the conviction that it may limit to contribute to the improvement of community. Artificial intelligence is utilized in different areas from improving instructive techniques to robotizing business measures. The growing thought of receiving AI in the medication improvement measure has moved from promotion to trust. In this survey, the conceivable utilization of AI in the medication improvement pipeline in medication improvement methodologies and cycles, the drug R&D productivity and wearing down, and associations among AI and drug organizations are talked about [1].

3.2 USE OF DEEP LEARNING/AI

The fuse of AI into the medical care organization succinctly appears in Figure 3.1[1]. Computer-based intelligence is portrayed as the utilization of strategies that empower PCs to emulate human conduct. AI likewise contains a subfield called machine learning (ML), which utilizes factual strategies with the capacity to learn with or without being unequivocally customized. ML is classified into regulated, unaided, and support learning. Regulated learning includes characterization and relapse strategies

FIGURE 3.1   Applications of drug discovery.

were the prescient model is created dependent on the information from info and yield sources. Yield from directed ML involves illness finding under the subgroup order, and medication adequacy and Absorption, Distribution, Metabolism, Elimination, Toxicity (ADMET) forecast under the subgroup relapse. Solo learning involves bunching and highlights-discovering techniques by gathering and deciphering information dependent on info information [2].

Through solo ML yields, for example, illness subtype disclosure from bunching and infection target disclosure from include discovering strategies can be accomplished. Fortification learning is to a great extent driven by dynamics in a given climate and execution to expand its exhibition. The yields from this kind of ML incorporate drug plans under dynamic and exploratory plans under execution where both can be accomplished through demonstration and quantum science [3]. A further subfield of ML called deep learning (DL) utilizes fake neural organizations that adjust also, gain from the immense measure of exploratory information. The large information and related information mining and calculation strategies could give us the ability to find new mixes that might be new medications, reveal or repurpose drugs that could be more powerful when utilized separately or in the mix, what's more, improve the region of customized medication dependent on hereditary markers. The development of DL was seen with the expanding measure of information and the consistent development of PC power [4].

The noticeable contrast that makes DL a subfield of AI is the adaptability in the design of neural organizations. For example, recurrent neural networks (RNNs) completely associate feed-forward organizations. It is accepted that, with the appropriate foundation of strategies in AI, we will observe the progress into a period of limited disappointments in clinical preliminaries and quicker, less expensive, and successful medication advancement measures [5].

3.3   DRUG DEVELOPMENT

The criticism driven medication advancement measure begins from existing outcomes acquired from different sources, for example, high-throughput compound and section screening, computational demonstrating and data accessible in the writing. This cycle substitutes enlistment and derivation. This inductive–deductive cycle, in the end, prompts upgraded hit and lead mixes. Computerization of explicit pieces of the cycle diminishes irregularity and blunders, and improves the proficiency of medication advancement. Once more plan strategies require information on natural science for *in silico* compound combination and virtual screening models that work as proxies for the biochemical and natural trial of adequacy and poisonousness [6].

Eventually, dynamic learning calculations permit the ID of latest with favourable exercises for the given illness target. The initial phase in medication advancement is the distinguishing proof of novel synthetic mixes with organic action. This organic movement can emerge from the cooperation of the compound with an explicit chemical or with a whole creature. The principal compound that shows action for the given natural objective is known as a hit. Hits are frequently found during the screening of synthetic libraries, PC reenactment, or screening of normally disengaged materials, for example, plants, microbes, and organisms. The distinguishing proof of a lead particle is the second step in medication advancement [7]. A lead is a substance compound that shows the promising potential that can lead to the advancement of another medication as a treatment for an illness. Distinguished hits are screened in cell-based measures prescient of the infection state and in creature models of illness to portray the viability of the compound and its plausible security profile [8].

When a lead compound has been discovered, its substance structure is utilized as a beginning stage for synthetic adjustments with the target of finding mixes with maximal remedial advantage and insignificant potential for hurt. During the cycle of lead age, hit atoms are methodically adjusted to improve their action and selectivity towards explicit organic targets while diminishing harmfulness and undesirable impacts. The synthetically related mixes inferred from a hit are called analogs and the cycle is alluded to as hit extension. Therapeutic scientific experts attempt to hit extension utilizing grounded natural science procedures. To increment the manufactured throughput, scientists center around a particular response or set of responses to collect structure impedes together to make an arrangement of analogs rapidly. A building block is a compound that has a responsive utilitarian gathering and iotas that communicate with the dynamic site of a natural objective [8]. This dynamic site is a particular area in the organic objective to which the compound ties through association powers. The authority of a substrate to a functioning site can be pictured as lock and key/initiated fit models [9].

3.4   EFFECTIVENESS OF RESEARCH AND RATE OF DEVELOPING THE DRUG

The drug-resemblance rules in medication advancement are carefully applied, yet drug organizations face impressive challenges in improving R&D productivity. R&D productivity is essentially a term used to depict the quantity of new medications endorsed by the agency to spent some amount on R&D alone [10]. The expense of finding and building up a medication has since heightened from US$ 800 million in the year 2008 to the current assessed figure of US$ 6 billion. The expense for drug advancement incorporates the total of disappointment; accordingly, the assessed cost is a normal gauge for another medication to be brought into the center [3]. The declining rate in the quantities of new medications affirmed per billion US$ spent on R&D is disturbing. The better than Beatles factor offers a relationship that portrays the far-fetched thought of building up another medication for a specific illness with pharmacological movement better than any of the existing endorsed drugs. The obstacles clarified here concern the destiny of each new medication that offers a superior remedial impact. With each new medication, the stakes are raised, and the R&D shortcoming happens here as it turns out to be increasingly more hard to clear the always advancing obstacles [11].

Careful controllers portray the expansion in severity and in the fixing of guidelines inferable from drug disasters previously. It is a reformist exertion to bring down the danger resistance in new medications and to make a guide for drugs with more secure profiles, which is suitable for the buyers in any case, extensively expands R&D use. This is trailed by the toss cash at it propensity, a tendency where R&D offices increment their enlistment in human and different assets with expectations of getting great quantifiable profits by being the first to dispatch another medication. This likewise adds to their high R&D uses. Finally, fundamental exploration beast power inclination alludes to overestimating the capacity of advances in fundamental exploration–particularly target recognizable proof and approval and savage power screening techniques in preclinical examination–to build the achievement rate in the clinical phases of medication improvement. What's more, it is realized that the helpful advantages of medications emerge from the associations with numerous proteins instead of a solitary protein [12].

Hence, it is trying to plan a medication that follows up on different targets. Another hypothesis proposed to clarify the decrease in R&D proficiency is the 'easy pickins' issue. This depicts the destiny of various manageable medication targets. The medications without any problem manageable medication targets were, at that point, imagined and that leaves us with the more troublesome medication targets. In spite of the energizing test for disclosure, endeavors to grow new medications will require a greater expense in R&D. The expansion in R&D costs just as high wearing down rates in the advancement cycle of another medication present difficulties to the drug business. The fundamental causes and potential measures to lessen steady loss rates have recently been checked [13]. The pace of whittling down was portrayed utilizing first-in-quite-a-while to register ten set up pharma organizations in the period 1991–2000, where the ordinary achievement pace of items that make it through turn of events and endorsement from the controllers is almost 11%. Almost 62% of new chemical entities (NCEs) in clinical preliminaries don't reach the center [4].

The major hidden wellsprings of weakening in late clinical stages are ascribed to clinical security and adequacy followed by a definition, pharmacokinetics and bio-availability, and harmfulness. The absence of viability that arose in helpful regions with high disappointment rates incorporates drugs for central nervous systems (CNS) and oncology applications, primarily owing to the absence of accessible fit-for-reason creature models in preclinical turn of events. Also, another contributing component for wearing down is the 'thin clinical examination' system embraced by drug organizations. This term alludes to the broad utilization of *in vitro* [14] and *in silico* models rather than creature models in the early phases of the medication improvement measure. The open doors for fortunate disclosure are additionally restricted with the current framework, since the current framework considers a medication competitor that applies other pharmacological impacts than the planned impact as a clinical preliminary disappointment [15]. Moreover, multicentre clinical preliminaries across the globe are additionally favored with the objective of getting different information. Be that as it may, this prompts daintily spread patients in which the specialists can pass up the occasion to notice new pharmacological impacts. Paradoxically, the clinical preliminary size for another subsequent medication in any helpful region is moderately high. This is on the grounds that the subsequent medication should be more powerful than the current medications for it to be qualified for enlistment [16] as in the Figure 3.2 [2].

According to the rules, the clinical preliminary size should be contrarily corresponding to the square of the distinction in the portion. For example, if the portion is split, the clinical preliminary size should be expanded multiple times. Every one of these imperatives add to an increment in expense of later-stage clinical preliminaries.

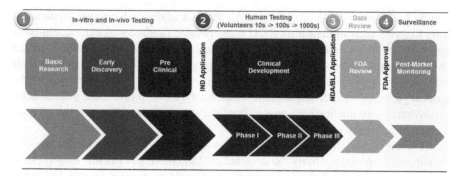

FIGURE 3.2 Overview of drug discovery.

The profit for the R&D venture is diminished by the accessibility of numerous treatment alternatives for a given illness, bringing about the low likelihood of causing another to catch up medication as a blockbuster drug [17]. Furthermore, the long process duration for endorsement inferable from the tough endorsement rules specified by the Food and Drug Administration (FDA) likewise goes about as a contributing element for a decrease consequently on R&D speculations. Nonetheless, these elements are not general as upheld by the logical writing that shows that diverse restorative zones have various paces of achievement [18]. For instance, biologics have a higher rate of accomplishment from first-in-quite-a-while to the facility, and licensing in mixes has generally higher achievement rates, at 24%. The pace of whittling down of accumulates with novel instruments of activity is higher than that of those with point of reference systems of activity. Pharma organizations can average out the diverse remedial zones of fluctuating achievement rates as a measure to adjust the, generally speaking, accomplishment to diminish the steady loss rates. In this way, by comprehension of their separate fundamental factors, the steady loss rates could be diminished [19].

3.5 DRUG DEVELOPMENT STAGES

Building up that medication, nonetheless, most likely takes longer than one would anticipate. By and large, getting a possible medication competitor from the research facility to the drug store takes around 14 years, costs more than US$ 1 billion, and has a minimum achievement rate. A fruitful medication will go through every one of the five phases: drug disclosure, pre-clinical exploration, clinical preliminaries, FDA endorsement, and post-market observing. Achievement measurements are assembled a few phases into drug advancement; another report states 13.8% of medication applicants that enter Phase I of clinical preliminaries, the first test in quite-a-while, will procure FDA endorsement. Upon endorsement, organizations have a restrictiveness period going from a half year to seven years to procure back the enormous cost of medication improvement before conventional prescriptions are delivered by contenders. Consequently, there is a fundamental financial drive related to drug organizations so they can proceed with the creation of life-saving medicine.

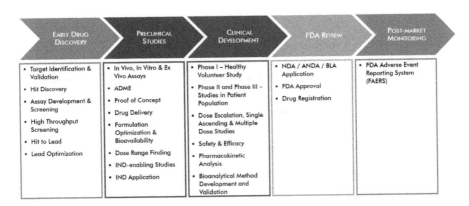

FIGURE 3.3   Stages of drug discovery.

There are five basic strides in the drug improvement measure, including numerous stages constantly inside every one of them. We will examine these various stages a lot to build up a top to bottom comprehension of the whole cycle. The five stages as in the Figure 3.3.

3.5.1 Drug Discovery

The fruitful medication with little results needs to interface with a particular objective, for example, a protein or quality. In any case, it is uncommon for an illness to have a known objective. To begin with, researchers need a comprehension of what drives the illness. Without this data, it is hard to anticipate an infection target and clarifies why researchers are as yet chipping away at solutions for illnesses. When atomic drivers are distinguished (outside of the medication revelation course of events), therapeutic scientists and researchers work through a progression of taught suppositions to ideally recognize an objective. In the event that an objective is known, researchers can screen through a huge number of mixes (for example a medication screen) to discover one that attractively influences the objective. Researchers deliberately select mixes to be utilized in the medication screen-based on their Structure Activity Relationship (SAR), in which Personal Computers (PCs) anticipate the natural impact based on the sub-atomic structure of the compound by contrasting it with comparative mixes with known impacts. It is imperative to take note of that, though all the more testing, researchers can at present play out a medication screen if the sickness target is obscure. For this situation, the medication screen is looking for aggravates that forestall certain qualities, or aggregates, of the infection.

On the off chance that the medication screen effectively recognizes mixes, researchers will direct trials with the mixes in cells developed on a plate (otherwise called *in vitro* explores) to guarantee the compound is compelling in the phone climate [20]. Sadly, drug revelation can become cyclic if the underlying compound has a helpless strength. An ineffectively powerful, nonselective compound may be updated, and hence rehash the medication screen measure. Distinguishing a promising medication could take a few endeavors more than 4–5 years. This stage likewise

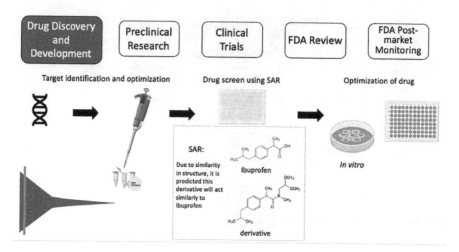

FIGURE 3.4   Drug discovery.

has the most minimal achievement rate, with just σ 250 out of 5,000 to 10,000 accumulates moving to the following round of testing. From Figure 3.4[3]. SAR recommends the subordinate will act also to ibuprofen and test in cells (*in vitro*) to additional test the compound.

3.5.2   PRE-CLINICAL STUDIES

At the point when a compound shows guarantee *in vitro* [14], a proposition is presented by the analyst to lead research inside a living creature, known as *in vivo*. The proposition is gotten by a board of trustees at the examination found that carefully controls creature exploration to guarantee the need and ethicality of analyses. On account of the physiological likenesses in people and creatures, creature models are significant in helping researchers in making an interpretation of their discoveries to people [20]. This preclinical information assists researchers with an understanding of the dose of the compound, results, contrasts among sexes, and critically, the viability of the compound in treating the specific sickness. The power from the medication screen in the medication advancement stage will give analysts a ballpark on what the measurement should be, yet that screen was acted *in vitro* [14]. The measurements study is essential in creatures to decide whether the medication has poison levels towards other cell types or organs, which would not be found in a test acted in cells [21]. Pre-clinical investigations take up to a year to accumulate information and generally just around five of the 250 mixes that entered pre-clinical examinations are effective.

A few reasons a compound may come up short incorporate high required dose, helpless results, or powerlessness to arrive at the objective, for example, going through the blood-brain barrier (BBB) [22]. Accordingly, it is an unfathomable accomplishment for a compound to have achievement in a creature show and get an endorsement to be tried in people through clinical preliminaries. The BBB is a layer between the mind and its encompassing veins. Even though it is astounding at

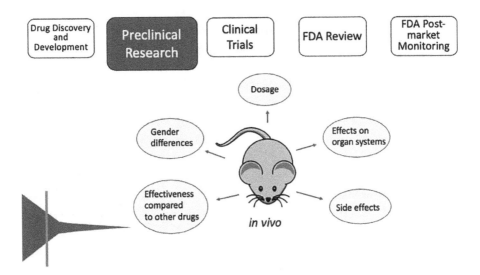

FIGURE 3.5 Pre-clinical studies.

shielding the cerebrum from sickness causing microbes and poisons in the blood, it additionally forestalls little atoms, similar to medicate up-and-comers, from arriving at the mind as in the Figure 3.5[3].

3.5.3 CLINICAL TRIALS

Exploration in people, much the same as in creatures, requires broad desk work and endorsement. There are three stages in clinical preliminaries, and as the compound moves to each stage, more individuals and checkpoints are added. Stage I enlists 10–50 patients to guarantee legitimate measurements in people. Since this is the first test in quite a while, extraordinary consideration is paid to notice any results; there is little worry of how the medication influences the infection as the essential center is if the medication is decent. In the event that the medication has too many results, it does not merit testing in wiped outpatients [23]. Appraisals from 2018 propose if a medication competitor passes Phase I, it has a 14% possibility of acquiring endorsement. Stage II selects many patients with the particular sickness they are planning to treat. Notwithstanding the possible medication, a fake treatment is actualized as a control. This aids researchers and specialists comprehend if their medication up-and-comer is really causing gainful changes and it isn't only a misleading impact [13].

Last is Phase III, adding up to around 1,000 patients selected and may take quite a long while to finish. Researchers are as yet paying special mind to results, critical contrasts in patient wellbeing contrasted with the fake treatment, and the new assignment of estimating if this new medication up-and-comer is superior to the norm of care, for example what is as of now used to treat those patients. Note that patients in clinical preliminaries will consistently get the fake treatment/drug competitor notwithstanding the norm of care. For instance, the norm of care for some tumors is a

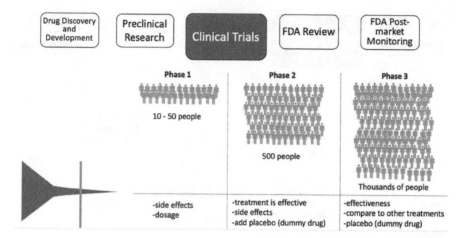

FIGURE 3.6   Clinical trials.

kind of chemotherapy [20]. In the event that there is a medication possibility for a specific malignancy, it is given in mix with chemotherapy. Be that as it may, if no standard treatment exists, patients will just get the test drug. Remember it is feasible for drug contenders to be dismissed anytime in these clinical preliminaries. For a medication to make it right to Phase III of clinical preliminaries is exceptional, and from here it has a half possibility of getting endorsed [23]. It is critical to take note that if a medication competitor demonstrates right off the bat that it was compelling for a perilous infection, the FDA will pronounce a quickened endorsement and give the medication to all enlisted patients as in the Figure 3.6[3].

3.5.4   FDA Analysis

Normally, just 14% of medications that are tried in people are submitted to the FDA for the survey. An application is submitted to the FDA, including all information that was created throughout the decade. Survey of a standard accommodation takes 10 months, yet an applicant that got need audit will be assessed in a half year. By that time, the FDA surveys all information, including the medication name, and investigates the office wherein the medication will be made [24]. During this assessment, officials review and notice records of the creation cycle and gather tests to secure human wellbeing. Upon endorsement, the FDA allows a restrictiveness period, or selective showcasing rights, going from a half year to seven years, contingent upon the classification into which the latest medication falls.

Selectiveness can run simultaneously with a patent, a property right given by the US Patent and Trademark Office that endures a long time from the date of utilization. In contrast to eliteness, a patent can be conceded anytime during drug improvement [20]. After the fruitful finish of Phase III clinical preliminary, an application is submitted to the FDA with all gathered information and a medication name. The FDA will assess the medication producing office to guarantee security as shown in the Figure 3.7[3].

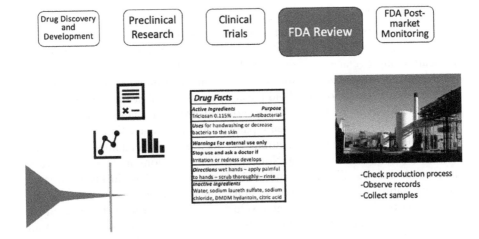

FIGURE 3.7 FDA reviews.

3.5.5 POST-MARKET REVIEW

In spite of the fact that the periods of clinical preliminaries Figure 3.8[3] are adequately long to screen any transient results, it is obscure how the medication will influence the body long haul. Stage IV of clinical preliminaries starts after the medication got endorsement to screen long haul impacts, notice if new results show up over the long haul, and decide whether the medication expands patients' life expectancies [20]. This stage is particularly significant for drugs that got quickened endorsement and consequently had more limited clinical preliminaries.

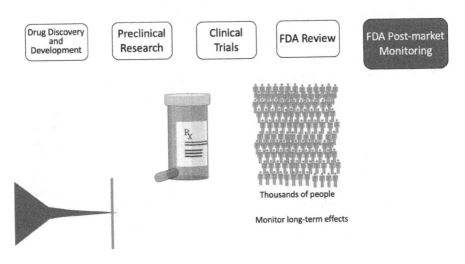

FIGURE 3.8 Post-market reviews.

3.5.5.1   The Society's Expense

A medication prevailing in FDA endorsement requires great expectations, arranging, the correct objective ID, and assets. The costs, dangers, and timetable of medication improvement are not regularly discussed, so justifiably, a patient who should pay an enormous total for an essential medication might be disturbed. In any case, these costs should uphold the advancement of that specific medication, just as the lost capital from recently bombed endeavors notwithstanding best endeavors [13][20]. Besides, when a patent and additionally time of restrictiveness terminates, different organizations can make and market a nonexclusive rendition of the FDA endorsed drug. Nonexclusive medications are less expensive on the grounds that drug disclosure and clinical preliminaries have just been directed; an organization delivering a conventional medication basically pays for the assembling of the pill once the nonexclusive is FDA affirmed. Because of the minimum price to purchasers, almost 80% of recommended drugs in the USA are generics [25]. Thus, the first maker of the medication would like to sufficiently offer to coordinate the expense of medication creation before the opposition starts.

Along these lines, there is a vital monetary drive related to large drug organizations; without cash and assets to start drug revelation, we wouldn't have any meds in any case. Moreover, in view of the significant expenses related to drug improvement, it is hard for drug organizations to explore treatment for uncommon infections. In the event that a drug organization fails to create a medication that helps a few hundred individuals, at that point millions are left without meds [24]. Apparently, there is a moral choice that should be made. This dilemma was acknowledged and in 1983, the Orphan Drug Act was passed by Congress to give monetary motivations to drug organizations to seek after medication improvement for uncommon illnesses. This demonstration prompted the making of 250 FDA affirmed vagrant medications. Moreover, there are instances of numerous people with a typical illness, yet don't have pay or protection to pay for medication. Six drug organizations, including Eisai, GlaxoSmithKline, etc., gave more than US$ 22.3 billion of medication to non-industrial nations to help battle this issue.

Despite the fact that medication costs for the buyer are huge, there is a requirement for huge scope drug improvement that takes numerous years and comes at a lofty cost. Likely approaches to reduce customer expenses could incorporate improving the initial step, drug revelation. As PC models improve, expectations of driving mixes may turn out to be more exact, bringing about minimum time and cash spent on the main stage. SAR, referenced prior, will help recognize likely components of activity of the new compound; these expectations might help in distinguishing results all the more rapidly [20]. Generally speaking, this likely cost-investment funds from improved computational models will likewise diminish the complete cost of medication advancement and ideally bring about less bombed mixes. Notwithstanding the measure of time and cash it takes to propel a medication from the lab to your medication bureau, all phases of medication improvement are important to guarantee wellbeing and to propel human wellbeing.

3.6   USE OF MACHINE LEARNING IN DRUG DISCOVERY

Drug discovery is generally viewed as the principal phase of a medication advancement pipeline and is an exploratory advance that points at revealing putative medication up-and-comers or quality targets, or causal factors, of a given sickness or a given synthetic compound. An assortment of deep learning techniques and machine learning techniques applied to biomedical issues have been altogether assessed in the last decade, with a developing revenue in Deep Learning (DL). For additional audit for DL strategies applied to sedate revelation, kindly allude to. Applications may handle intriguing issues with regards to tranquilize disclosure: for instance, drug up-and-comer ID through particle docking, to anticipate and pre-select fascinating medication target collaborations for additional exploration; and protein designing, that is, all over again the atomic plan of proteins with explicit anticipated official or theme capacities.

A genuinely ongoing achievement in protein configuration utilizes generative DL, all the more accurately, Generative Adversarial Networks (GANs). A GAN is made of two at the same time prepared neural networks (NNs) with particular jobs: a Generator, which is prepared to test occasions, and a Discriminator. The objective of the latter is to perceive preparing occasions from produced ones, by allotting them a likelihood estimation of the thought about the case being tested from the preparation set. The goal is to prepare the Generator to make counterfeit occasions that can fool the Discriminator, that is, which are persuading enough regarding the fundamental work done to by the Discriminator. Figure 3.9 [26] which represents a GAN. For

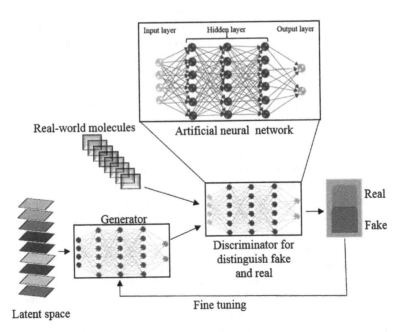

FIGURE 3.9   GANs [26].

example, GANs have been applied to create DNA successions coordinating explicit DNA themes, that is, short DNA grouping variations that are related to a particular capacity. As a proof-of-idea, exceptionally appraised tests got by means of this methodology, showing one or a few duplicates of the ideal theme, were appeared. Another way of utilizing the NN which predicts the likelihood of the DNA arrangement of coding for an antimicrobial peptide (AMP), furthermore, has prevailed with regards to preparing a GAN based on time AMP-coding successions.

Drug discovery issues have likewise persuaded research in blackbox enhancement, generally utilizing Bayesian Optimization (BO), see, for example, the accompanying referred to models. BO is a field of exploration for finding worldwide ideals of a discovery work by consecutively choosing where to assess this capacity. For sure, the alleged discovery target work is open as it were through its qualities at chose focuses and may be expensive to assess. The premium in a successive methodology is to adaptively pick where to gather data next. The inspiration for review some medication disclosure assignments as a discovery enhancement issue comes from the way that they typically depend on costly reenactments, for occasion, protein-collapsing, making programmed drug screening, what's more, evaluation by means of these tasks tedious. The particularity of BO is to pick an earlier likelihood dispersion on the goal work: e.g., one accepts that the target work is a test of a given likelihood conveyance over capacities. This is then refreshed after each new capacity assessment into a back appropriation, which, in turn, controls the choice of the new highlight assessment.

For example, the creators apply BO to plan quality arrangements that boost record and interpretation rates, from beginning groupings. The streamlining is performed on element vectors of fixed length related to the quality groupings, and the target work is the capacity which, given a quality grouping highlight vector, restores the related record and interpretation rate capacities. The earlier upon the target work is a Gaussian Process, traditionally utilized in BO. The creators at that point use as obtaining capacity the normal goal, which will boost the normal of the record and interpretation rates. The highlight vector amplifying this securing capacity will, at that point, characterize an ideal quality plan rule for the augmentation of both records, furthermore, interpretation rates. Since a few codons can code for the equivalent amino corrosive, given a bunch of successions coding for a protein of interest, these successions can be positioned by the comparability of their relating highlight vector with the ideal quality plan rules that have been inferred. The thought is to deliver then the protein at a premium with lower costs since the protein creation rate is boosted. Another illustration of uses of BO to tranquilize revelation is to invigorate the disclosure of new substance mixes, that is, discovering little atoms that may advance for a property of interest, while being artificially not quite the same as known accumulates.

ML calculations have essentially progressed drug disclosure. Drug organizations have extraordinarily profited by the usage of different ML calculations in medication disclosure. ML calculations have been utilized to create different models for foreseeing compound, organic, and physical qualities of mixes in medication disclosure. ML calculations can be joined on the whole steps of the cycle of medication disclosure. For instance, ML calculations have been utilized to locate another utilization of medications, foresee drug-protein connections, find drug viability, guarantee security biomarkers, and improve the bioactivity of atoms. ML calculations that have been generally utilized in medication revelation, which include: Random Forest (RF), Naive Bayes (NB), and support vector machine (SVM) just as different strategies.

ML calculations and procedures are not a solid, homogeneous subset of AI. There are two fundamental kinds of ML calculations: Supervised and unsupervised learning. Unsupervised learning learns from preparing tests with realized names to decide the names of new examples. Unsupervised learning perceives designs in a bunch of tests, ordinarily without names for the examples. The information is, as a rule, changed into a lower measurement to perceive designs in high-dimensional information utilizing unaided learning calculations before perceiving designs. Measurement decrease is valuable, not just, on the grounds that unaided learning is more effective in a low measurement space yet additionally in light of the fact that the perceived example can be all the more effectively deciphered. Regulated and unaided learning can be consolidated as semi-regulated and support realizing, where the two capacities can be used for different informational indexes.

Sweeping volumes of information are basic for the turn of events, advancement, and reasonability of effective ML calculations in each progression of the medication disclosure measure. The dependence on enormous great information and known, very much characterized preparing sets are significantly more fundamental inaccuracy medication and treatments inside medication disclosure. Accuracy medication requires an exhaustive portrayal of all connected container omic information: Genomic, transcriptomic, proteomic, and so on, to aid growing really successful customized drugs. The far-reaching utilization of high-throughput screening and sequencing, online multi-omic information bases, and ML calculations, in the previous 20 years, have established a thriving climate for some parts of the information age, assortment, and support needed for drug advancement.

The progressions of information examination have effectively endeavored to depict and decipher the created information. This undertaking, upheld with ML procedures and incorporated information bases through various programming/web-instruments as shown in Tables 3.1[4] and 3.2[4], is currently routinely utilized for all means in medication disclosure. The capacity of new information examination to synergize with old-style draws near and earlier theories to create novel speculations and models has substantiated itself to be valuable in uses of repositioning, target disclosure, little particle revelation, union. The overall strides in medication

TABLE 3.1
Tools/Software[4]

Tools/Software	Purpose
PharmMapper	Drug target identification
Drug Target Ontology	Text Mining, Classification
MANTRA	Network Analyzer
ABNER	Text Analyzer
GeneWays	Extracting Tool.

TABLE 3.2

Database[4]

Database	Purpose
BindingDB	medicinal chemistry
BioGRID	Provides network support
ChEMBL	Keyword searching
CARLSBAD	Associativity identification

revelation. AI and deep learning algorithms may take an interest in every one of the four stages recorded, e.g., by mining proteomic in objective revelation, finding little particles as competitors in lead disclosure, creating quantitative structure-action relationship models to streamline lead structures for improved bioactivity, and breaking down monstrous measure results.

3.7 TO TEST THE DRUG

When one or a few medication competitors are chosen, preclinical and clinical turn of events start. Drug properties, identified with body preparation of the competitor particle, should be surveyed in beginning stages: e.g., the ADMET properties: assimilation, dissemination, digestion, and discharge, alongside the harmfulness levels. Assessment of their viability is performed later during the computerization and improvement of *in silico* forecast models may get a good deal on later testing stages and resulting in *in vitro* and *in vivo* tries. In addition, these strategies may come to supplant probes creature models, since certain organizations begin prohibiting creature testing. Such examples of potential *in silico* directing of wet-lab investigations can be found in. For example, a chart-based structure is created in request to assemble a forecast model for bother tests, given time-arrangement articulation information and putative quality administrative cooperations between qualities of interest. The subsequent model can at that point foresee articulation levels for the chose set of qualities.

This technique may help evaluating the medication consequences for pathways after bother of the medication targets. The creators have approved their strategy on a mouse pluripotency model, within *in vitro* examinations, and report that 61% of the anticipated aggregates could be replicated *in vitro*. Additionally, given the inborn consecutive nature of a clinical preliminary, in which patients are given medicines, one patient after the other, calculations would be characteristic contender to be utilized in additional periods of medication testing. Nonetheless, these days, the inspiration for growing new crook calculations has altogether moved to applications to online substance enhancement, for example, successive recommender frameworks. Criminal plans appear to have been seldomly utilized in human clinical preliminaries. Conventional randomized clinical preliminary (RCT) has been the highest quality level. In any case, the previous years have seen an expanded interest in all sorts of Adaptive Clinical Trials (ACT), in which the following assigned treatment could be needy of the result of recently dispensed medicines. We will presently expound on

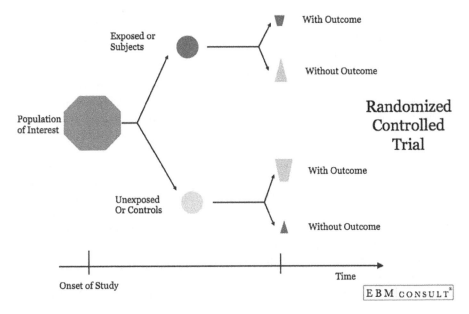

FIGURE 3.10   RCT.

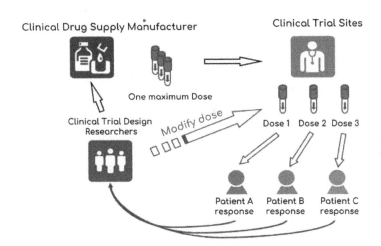

FIGURE 3.11   ACT.

the last mentioned. Figures 3.10[5] and 3.11[6] presents the scenario of the RCT and ACT, respectively.

This adaptivity could prompt a more modest example size to accomplish a given force, or early halting for harmfulness or purposelessness of treatment, via consequently performing break investigations. The issue of augmenting prizes in such a criminal model was widely concentrated from the 1950s, either from the achievement of probabilities that are treated as obscure boundaries to be construed; or from a Bayesian view where achievement probabilities are accepted to come from some earlier conveyance, likewise to Bayesian advancement. Augmenting rewards adds up to expanding the number of relieved patients, which is not the motivation behind a clinical preliminary. However, the factual network has likewise taken a gander at the distinctive issue of finding, as fast and precisely as could be expected under the circumstances, the treatment with the biggest likelihood of achievement. This was read, for instance, under the name positioning and determination and later best arm distinguishing proof. A fascinating take-out from the crook writing is that the two destinations of treatment recognizable proof and relieving patients can't be accomplished by a similar designation procedure.

In any case, as appealing as deciding the best treatment while treating appropriately however many patients as could reasonably be expected is the utilization of ACTs remains highly uncommon. This may be because of inherent contrasts between conventional clinical preliminary system and information subordinate distribution. For instance, balance between prognostic covariates in each treated gathering of patients is needed for factual significance. The utilization of ACTs may likewise be frustrated by pragmatic reasons in the clinical preliminary setting, for example, when one should manage significantly postponed input. In any case, these downsides may be relieved by the advantage hazard proportion of the treatment. For life threatening illnesses, a versatile clinical preliminary may be an expectation for the patients to improve their condition, for best observational medicines, utilizing past perceptions, can be appointed to patients. Notwithstanding this underlying aggression toward ACTs, there has been a new flood of interest in advancing their genuine use. As a striking sign of this development, the FDA has refreshed a draft of rules concerning versatile clinical preliminaries, posting specifically a few concrete instances of effective versatile preliminaries. Essentially, the method presents a few models and advances some great practice for utilizing ACTs.

Meanwhile, the creators have played out numerous recreations delineating the attributes of normal desperado calculations as far as allotment and last determination, to promote their utilization. Among existing criminal calculations, Bayesian calculations have accomplished a specific prominence. Strangely, the absolute first desperado calculation can be followed back to crafted by Thompson in 1933, who proposes randomizing the medicines as indicated by their back likelihood of being ideal. This standard presently called Thompson Sampling or back examining was rediscovered around 10 years prior in the ML people group for its phenomenal exact execution in complex models. Strangely, the utilization of variations of this guideline seems to have been proposed too in various setting of medication testing. For instance, different preliminaries present a trade-off between RCT and Thompson Sampling, while presenting a proof-of-idea for the Alzheimer's sickness that depends on such back inspecting thoughts. All the more extensively, a few instances of effective Bayesian versatile plans have arisen throughout the most recent 20 years, and we allude the peruser for a study.

3.8 DRUG REPURPOSING

The difficulties of planning new atomic substances, and testing them through every single clinical stage, have created research interest in a more beneficial and proficient procedure, called drug repurposing, or on the other hand drug repositioning. This methodology targets concentrating as of now accessible medications and synthetic mixes to discover new restorative signs. This technique is valuable when repurposed drugs have all-around recorded wellbeing profiles that is, results and medicines are known, and known instruments of activity. Various methodologies have been utilized to handle the medication repurposing issue. For instance, some depend on programmed preparation of Electronic Health Records (EHR), clinical preliminary information, and text mining techniques to recognize relationships between's medication particles and quality or protein focuses on writing. Nonetheless, this approach may be delicate, yet not generally explicit, since the text understanding is as yet a difficult issue, and the connection between infection factors and medications probably won't be clear.

The present status of the workmanship techniques appear to have gone to various standards of repurposing, see for example, the accompanying surveys for an order of these various strategies. In any case, a large portion of these strategies depends on solid speculation, which is that closeness between components–for the occasion, the compound arrangement of medication particles–suggests connection at remedial impact level, or at the drug target level. Regardless, counter-guides to this speculation have been appeared to prompt awful occasions: for example, thalidomide exists as two chiral structures. One of these structures can treat morning ailments; the other structure can have teratogen impacts. An endeavor to measure all the more precisely drug impacts is the mark inversion, likewise called availability planning, which centers around articulation estimations: given a neurotic aggregate related with the illness at study, the goal is to distinguish which medicines are generally ready to return this mark.

This activity is performed thorough examinations of the questioned signature with purported drug marks, that is, vectorized outlines of gene-wise articulation changes because of the thought about medications. This approach has recorded a few achievements in medication repurposing; for example, in an extensive survey of this kind of strategy. Nonetheless, note that depending just on transcriptomic estimations to comprehend the component behind an illness may lead to wrong headings when these can't represent its causal factors. Also, drug repurosing methods that straightforwardly use drug marks separated from the Library of Integrated Network-Based Cellular Signatures (LINCS) information base, either to register a closeness measure or to foresee treatment impacts at the transcriptome level, are one-sided: without a doubt, the medication marks that have been registered in LINCS are estimated in harmful or deified or pluripotent cells–transcriptomic guideline might vary from solid cells. Moreover, particularly in grid factorization and some deep learning strategies that play out a measurement decrease or, more, by and large, highlight learning on the medication marks, the educated highlights regularly barely bode well naturally, which forestalls simple translation of the outcomes and once-overs to verify everything seems good.

Another method of tackling the medication repurposing issue is to consider it to be a recommender framework issue, where a specialist should suggest the best accessible choices. Here, the specialist should choose the most encouraging medication competitors with regard to the illness or the objective of the study. Some new papers have received this methodology: for example, the creators have planned a diagram-based technique to anticipate drug target or drug-illness connections. Given a medication, the model predicts a rundown of fixed length, that contains illness-related targets in all likelihood influenced by the substance compound.

3.9 DIFFICULTIES FACED BY MACHINE LEARNING IN DRUG DISCOVERY

Presently, the vital difficulties in medication revelation are identified with synthetic structures, picking the correct medication target, or the fitting measurements for a particular patient subpopulation. In spite of significant advances in clinical pharmacology, the storehouse ing of late preclinical and clinical information stays a basic issue impeding the organization of AI advancements. The drug business doesn't regularly share pharmacokinetic and pharmacodynamic estimations for the vast majority of the competitor drugs or their mixes, except if the drugs are affirmed for human use. Undoubtedly, not at all like different zones of innovative work, just a little portion of the medication disclosure information generally is accessible for preparing ML models. This is especially as for genuine negative information.

The issue infests not just for, for instance, separating drugs from non-drugs, yet in addition concerning ML models for novel objective illness relationship, for understanding why clinical preliminaries were suspended, why medications were removed, or in any event, getting to finish datasets from fruitful clinical preliminaries. In any case, there is presumably that clinical improvement is at present traveling through significant change and that the use of ML is quickening. Consistently, a clinical pharmacologist is a therapeutically qualified expert who educates, explores, outlines strategy, and offers data and guidance regarding the activities and legitimate employments of drugs in people, and actualizes that information in clinical practice. We can securely express that new AI/ML advancements are probably going to affect clinical pharmacologists at all levels in the following: (a) In educating, the utilization of information escalated techniques for instance, writing looking and preparing, just as communications with on-line prescient ML models, are probably going to increment.

Comprehensive assets, for example, ADMETLab, or explicit assets, for example, programming for drug-drug interactions are probably going to profit and co-develop with the expanded utilization of AI/ML techniques. (b) In the examination, the utilization of AI/ML techniques is quickening pace, as defined above. (c) In outlining strategy, AI/ML techniques are foreseen to impact medical services in many nations. (d) For the legitimate utilization of prescriptions, most advanced locals (i.e., patients and researchers who began to utilize PCs or tablets or cell phones since the beginning) are looking for medical care and drug data by means of online media and electronic stages. Present investigation (e.g., foreseeing inquiry terms and customized looks) for drug-related data are now accessible. Progressed AI/ML frameworks, custom fitted to a person's clinical and genomic history are around the corner.

One of the significant difficulties of the drug business is that the numerous regions of medication disclosure and improvement can take years, cost countless dollars, and are frequently detached from the arranging point of view and oversaw by various individuals. The time it takes to recognize an infection target and figure an organic speculation may take many years and cost billions. The commonplace time from recognizing a hit particle for an offered focus to an advertised item normally takes 12.5 years. What is more, despite the fact that the sum of clinical pharmacology information is quickly expanding, it is probably going to be significantly less plentiful than the measure of test science and *in vitro* high-throughput screening information. There are not many investigations with estimating countless boundaries *in vitro*, in mice, and in people with distributed openly accessible information.

Most of these examinations were directed by the drug organizations that legitimately treat this information as an upper hand, making clinical pharmacology more difficult for the advances in AI/ML to upset. The following decade may observe consistent incorporation between human knowledge and AI frameworks. Astute individual aides effectively offer help regarding the route, amusement, and shopping. Clinical counsel, explicitly with respect to the utilization of drugs, is highly likely on the to-do list for most engineers contending in this area. It stays to be seen which one will be supported by proficient affiliations, for example, the American Society for Clinical Pharmacology and Therapeutics, based on exactness and accuracy. Albeit most AI/ML improvements are expected to profit patients and researchers the same, we do foresee that these patterns won't supplant clinical pharmacologists in one decade from now, and clinical pharmacology by and large will be disturbed more slow than different fields of medication disclosure and advancement.

There is, be that as it may, a solid impetus for clinical pharmacologists to adjust, to screen such patterns, and to get insightful on the utilization and dangers of AI/ML frameworks, especially for those that impact their everyday practice. There should be a network and administrative exertion for open information sharing to make bigger number of preclinical and clinical pharmacology information accessible for the ML network to use to quicken the AI/ML progress in the territory.

3.10   DISCUSSION

Recent advances in AI, including the development and improvement of more unpredictable ML procedures, have had a tremendous effect on the drug advancement measure. Artificial intelligence innovation can address a portion of the fundamental difficulties strikingly diminishing cost, time, and work requests during the beginning phases of medication disclosure, by misusing *in silico* approaches for a new drug plan, combination forecast, and bioactivity forecast. An expansion procedure from ML is DL. In this article, we have discussed the hypothetical establishments and praiseworthy late utilization of DL in chemoinformatics and drug disclosure. The DL advances are a stride in front of ML innovations. Customary ML methods depend on profoundly carefully assembled highlights in extraction or designing. For example, to get nice outcomes in picture arrangement, a few preprocessing strategies must be applied, for example, channels, edge identification, etc. The greatest preferred position of DL is that most, if not everything, highlights can be educated naturally from the information, given that enough preparing information models are accessible. DL models have highlight identifier units at each layer that slowly extricate more mind-boggling and invariant highlights from the first crude information signals. Lower layers distinguish straightforward highlights that are then taken care of into higher layers, which thus recognize more mind-boggling highlights.

In contrast, conventional ML models present not many layers that map the first information highlights into an issue explicit component space. Traditional ML approaches to tackle the issue by separating the issue, addressing various parts first, and afterward joining the outcomes at long last to give yield, while DL approaches to tackle the issue utilizing a start to finish approach, that is, straightforwardly taking care of the data sources and comparing yields, without determining any standards or examples to the organizations and without deteriorating encoding preparing and translating preparing into two separate advances. In contrast to old-style ML structures, DL models set aside long effort to get prepared in view of the colossal number of boundaries and generally gigantic datasets, which restricts its application for the most part. For instance, on the off chance that one has just a modest quantity of action information accessible, it is hard to anticipate the organic action of the new particles in light of the fact that a couple of information can't cover an adequate compound space.

The one-shot learning strategy, which requires just restricted preparation tests, can be utilized in tackling such an issue. At long last, regular ML techniques are very prohibitive, though DL strategies can be applied to a wide scope of utilizations and different areas. An enormous piece of it goes to the exchange discovering that permits making use of pre-prepared DL networks for various errands inside the equivalent spaces. In spite of the fact that we are examining here all the preferences about DL over the other ordinary ML approaches, we can't state that DL is better than the customary ML strategies. There is no single calculation that can address all ML issues more effectively than others. Everything relies upon explicit application cases. For instance, when the issue includes joined arrangements of a compound or target protein input descriptors, the execution of DL doesn't fundamentally vary from other ordinary ML calculations. On the other hand, DL has indicated much better execution in biomedical picture bunching and examination just as in foreseeing organic movement and harmfulness, compound responses what's more, again sub-atomic plan. Therefore, it very well may be derived that DL along with ML can make colossal advances cooperatively in the improvement of medication disclosure viewpoints.

However, the huge advancement has been made in the field of DL application in drug revelation, it remains a field with numerous huge challenges, for example, the reliance on utilizing test information for preparing and approval, and the scoring of mixes official to proteins. Generous headway is being made in the field of move learning, one-shot learning, and conformal forecast, with ongoing upgrades in free-energy, which adds to addressing the scoring challenge. Since the preparation of profound models with countless free boundaries speaks to a perplexing improvement issue, numerous research works have been dedicated to the advancement of proficient learning strategies for profound organizations. The techniques proposed to tackle issues of preparing profound structures to incorporate the improvement of more effective analyzers, the utilization of initiation capacities dependent on nearby rivalry, the utilization of all-around planned in statement strategies, and the utilization of skip associations between layers with the point of improving the progression of data.

Notwithstanding, the deep learning method still has a few issues caused by the stacking of a few non-direct changes that need to be tended to. Further investigation into the age of new descriptors just as the improvement of advantageous rules for a right portrayal of a compound framework will be basic for the not so distant future. The accessibility and versatility of DL models can quicken future improvements by means of learned highlights and hypothesis educated models. Besides, since our human psyche isn't reliable and we don't have the information to test the achievement of restorative science programs, we should do whatever it takes not to set the benchmark for DL techniques excessively high. At last, to take the full favorable position of AI procedures, later on, critical assets will be required for information care, coordination, and executives. The test affirmation of the AI viability in medication disclosure programs is an essential factor in the agreement on how AI can add to restorative science and how it very well may be enhanced and improved.

3.11 CONCLUSION

The making of new computerized stages and programming can impressively build ML development in a specific way. With regards to DL, the main future bearing is to consolidate the traditional profound structures with the expect to accomplish better and more elevated levels of joining between science information, biomedical information, omics information, and hypothetical calculation about the atomic elements. At last, it prospects a future pattern in the utilization of quantum ML for biochemical endeavors included in the beginning stages of medication disclosed. Quantum ML, which sits at the convergence of quantum material science and ML is arising as an amazing system permitting quantum speed-ups and improving old-style ML calculations. Therefore, the field is available for genuine investigations of how synthetic wave work information can be utilized by quantum calculations for solid medication disclosure applications.

CONFLICT OF INTEREST

There is no conflict of interest.

NOTES

1   https://igniteoutsourcing.com/healthcare/artificial-intelligence-in-healthcare/.
2   https://www.nebiolab.com/drug-discovery-and-development-process/.
3   https://misciwriters.com/2019/11/15/from-the-lab-to-your-medicine-cabinet-a-timeline-of-drug-development/.
4   https://www.biopharmatrend.com/post/45-27-web-resources-for-target-hunting-in-drug-discovery/.
5   https://www.ebmconsult.com/articles/randomized-controlled-trial-rct.
6   https://www.jliedu.com/blog/adaptive-design-clinical-trials/.

REFERENCES

[1] Y. Chen, and J. Kirchmair, "Cheminformatics in natural product-based drug discovery," *Molecular Informatics*, vol. 39, no. 12, pp. 2000171, 2020.

[2] I.I. Baskin, "The power of deep learning to ligand-based novel drug discovery," *Expert Opinion on Drug Discovery*, vol. 15, no. 7,pp. 1–10, 2020.

[3] D.G. Cheirdaris, "Artificial neural networks in computer-aided drug design: An overview of recent advances," in *GeNeDis 2018*. 1em plus 0.5em minus 0.4em Springer, 2020, pp. 115–125.

[4] F. Gentile, V. Agrawal, M. Hsing, A.-T. Ton, F. Ban, U. Norinder, M.E. Gleave, and A. Cherkasov, "Deep Docking: A Deep Learning Platform for Augmentation of Structure based Drug Discovery," *ACS Central Science*, vol. 6, no. 6, pp. 939–949, 2020.

[5] J.E. Kelly, C. Chrissian, and R.E. Stark, "Tailoring NMR experiments for structural characterization of amorphous biological solids: A practical guide," *Solid State Nuclear Magnetic Resonance*, vol. 109, pp. 101686, 2020.

[6] R.M. LeBlanc, and M.F. Mesleh, "A drug discovery toolbox for nuclear magnetic resonance (NMR) characterization of ligands and their targets," *Drug Discovery Today: Technologies*, 2020.

[7] X. Lin, X. Li, and X. Lin, "A review on applications of computational methods in drug screening and design," *Molecules*, vol. 25, no. 6, pp. 1375, 2020.

[8] P. Schneider, W.P. Walters, A.T. Plowright, N. Sieroka, J. Listgarten, R.A. Goodnow, J. Fisher, J.M. Jansen, J.S. Duca, T.S. Rush et al., "Rethinking drug design in the artificial intelligence era," *Nature Reviews Drug Discovery*, vol. 19, no. 5, pp. 353–364, 2020.

[9] C.S. Tautermann, "Current and future challenges in modern drug discovery," in *Quantum Mechanics in Drug Discovery*. 1em plus 0.5em minus 0.4em Springer, 2020, pp. 1–17.

[10] S. Mishra, S.R. Pandey, D. Hicks, A. Goyal, and D. Gaurav, "A blood pressure and heartbeat anomaly detection and notification mobile application system," *Journal of Web Engineering*, vol. 19, no. 5-6, pp. 735–761, 2020.

[11] J.J. Lim, J. Goh, M.B.M.A. Rashid, and E.K.-H. Chow, "Maximizing efficiency of artificial intelligence-driven drug combination optimization through minimal resolution experimental design," *Advanced Therapeutics*, vol. 3, no. 4, pp. 1900122, 2020.

[12] D. Gaurav, S.M. Tiwari, A. Goyal, N. Gandhi, and A. Abraham, "Machine intelligence-based algorithms for spam filtering on document labeling," *Soft Computing*, vol. 24, no. 13, pp. 9625–9638, 2020.

[13] S.P. Chatrati, G. Hossain, A. Goyal, A. Bhan, S. Bhattacharya, D. Gaurav, and S.M. Tiwari, "Smart home health monitoring system for predicting type 2 diabetes and hypertension," *Journal of King Saud University-Computer and Information Sciences*, 2020.

[14] M.J. Moreno, B. Ling, and D.B. Stanimirovic, "In vivo near-infrared fluorescent optical imaging for cns drug discovery," *Expert Opinion on Drug Discovery*, pp. 1–13, 2020.

[15] S. Tiwari, and A. Abraham, "Semantic assessment of smart healthcare ontology," *International Journal of Web Information Systems*, vol. 16, no. 4, pp. 475–491, 2020.

[16] A. Zhavoronkov, Q. Vanhaelen, and T.I. Oprea, "Will artificial intelligence for drug discovery impact clinical pharmacology?" *Clinical Pharmacology & Therapeutics*, vol. 107, no. 4, pp. 780–785, 2020.

[17] D. Gaurav, S. Shandilya, S. Tiwari, and A. Goyal, "*A machine learning method for recognizing invasive content in memes*," in *Iberoamerican Knowledge Graphs and Semantic Web Conference*. 1em plus 0.5em minus 0.4em Springer, 2020, pp. 195–213.

[18] Y. Zhou, F. Wang, J. Tang, R. Nussinov, and F. Cheng, "Artificial intelligence in covid-19 drug repurposing," *The Lancet Digital Health*, 2020.

[19] S.M. Tiwari, S. Jain, A. Abraham, and S. Shandilya, "Secure semantic smart healthcare (S3HC)," *Journal of Web Engineering*, vol. 17, no. 8, pp. 617–646, 2018.

[20] C.N. Cavasotto, and J.I. Di Filippo, "Artificial intelligence in the early stages of drug discovery," *Archives of Biochemistry and Biophysics*, pp. 108730, 2020.

[21] S. Keiffer, M.G. Carneiro, J. Hollander, M. Kobayashi, D. Pogoryelev, A. Eiso, S. Theisgen, G. Müller, and G. Siegal, "NMR in target driven drug discovery: why not?" *Journal of Biomolecular NMR*, pp. 1–9, 2020.

[22] C.S. Lee, and K.W. Leong, "Advances in microphysiological blood-brain barrier (bbb) models towards drug delivery," *Current Opinion in Biotechnology*, vol. 66, pp. 78–87, 2020.

[23] A. Moskalev, "Is anti-ageing drug discovery becoming a reality?," *Expert Opin Drug Discovery,* vol. 15, no. 2, pp. 135–138, 2020.

[24] E.G. Cleary, M. Jackson, A. Acevedo, and F.D. Ledley, "Characterizing the public sector contribution to drug discovery and development: the role of government as a first investor," *Institute for New Economic Thinking*, 2020.

[25] D.R. Flower, "Drug discovery: Today and tomorrow," *Bioinformation*, vol. 16, no. 1, pp. 1, 2020.

[26] M. Batool, B. Ahmad, and S. Choi, "A structure-based drug discovery paradigm," *International Journal of Molecular Sciences*, vol. 20, no. 11, pp. 2783, 2019.

4 A Detailed Comparison of Deep Neural Networks for Diagnosis of COVID-19

M.B. Bicer and Onur Dogan
Izmir Bakircay University, Turkey

O.F. Gurcan
Sivas Cumhuriyet University, Turkey

CONTENTS

4.1   INTRODUCTION

In December 2019, the latest pandemic was first encountered in Wuhan, China. It was called a disease resulting from a family of viruses named "Severe Acute Respiratory Syndrome Coronavirus 2" (SARS-CoV-2). Then the World Health Organization (WHO) called it COVID-19 in February 2020. Only seven species are hazardous to people from dozens of viruses that exist in the corona family (Coronaviruses, 2020). However, due to the fast growth rate of the COVID-19 problems, many developed countries' health systems have faced struggles managing the pandemic. Many governments have announced lockdown and demanded their citizens stay at home and rigidly avoid gatherings. It is declared that coronaviruses are transferred to individuals via mammals like bats (Fan et al. 2019). Originally, Chinese

DOI: 10.1201/9781003161233-4

citizens in Wuhan became infected by a coronavirus in December 2019, and the current has outbroken (Iqbal et al. 2020; Siddiqui et al., 2020). An infected patient may have several symptoms and signs, such as fever, dyspnea, cough, and headache. In some critical cases, the virus may result in pneumonia, multi-organ failure, and even death (Mahase, 2020). At the time of the study, any particular drug or vaccine has not been developed. Because of the sudden occurrence and mortal effects of COVID-19, many scientists and research centers have actively focused on achieving an efficient diagnostic way and prevention by a vaccine for the pandemic medication (WHO, 2020). These research topics consider not only medical or biotechnology technologies but also data-based solutions that apply artificial intelligence (AI) and machine learning (ML) to avoid and manage the pandemic by contributing technical and scientific insights and presenting potential solutions (Butt et al., 2020; Gozes et al., 2020; Hall et al., 2020).

A crucial action in challenging the pandemic is to screen the infected people so that they can be isolated and treated. Real-time reverse transcription-polymerase chain reaction (rRT-PCR) is one of the principal techniques to screen COVID-19 patients (Corman et al., 2020; Wang et al., 2020b). The rRT-PCR method gives the results in a few hours to two days by implementing on patient's respiratory samples. Chest radiography images can be useful to identify COVID-19 as an alternative of respiratory samples (Bernheim et al., 2020; Xie et al., 2020). A chest radiology-based system can be a practical tool in detecting and following up COVID-19 cases. The differences between infected and healthy people's lung images are compared to identify whether a person has COVID-19 (Fang et al., 2020; Xie et al., 2020). The identification system of COVID-19 using chest radiology images can be more beneficial than traditional ways. It can give rapid results and examine concurrently in many cases. These systems can be advantageous in health centers with a few numbers of testing kits (Khan, Shah & Bhat, 2020). Moreover, because radiology imaging systems are common in all hospitals, radiography-based approaches are more accessible.

The fact that a recovered patient may turn positive in a diagnostic test performed after 5–13 days demonstrates that even recovered patients may spread the virus (Lan et al., 2020). Hence, the accuracy of applied methods has more importance in COVID-19 diagnosis. Because analysis using X-ray images is not efficient in detecting the early stages of COVID-19 (Zu et al., 2020), a well-trained deep learning (DL) model can fill the deficiency that cannot be realized at the first glance. Moreover, the fast growth of the COVID-19 cases has required an automated identification and diagnosis system, AI approaches have been implemented in many types of research. Because deep learning methods do not depend on standard handcrafted characteristics, they have been applied in numerous image analysis (Ardakani et al., 2020; Panwar et al., 2020; Zhu et al., 2020). Convolutional Neural Networks (CNN), one of the mostly applied deep learning classifiers, has been dominant in medical image analysis by mapping lung images. Because coronavirus targets the lungs, analyzing the changes in the lungs with radiography images can give a precise result of the virus's presence.

In this study, deep learning models are compared for the diagnosis of COVID-19 in a faster and more accurate way by detecting some specific visual marks for coronaviruses such as ground-glass opacity and dark spots. All AI models applied in this study first extract features by CNN, then use various classifiers. The remaining part of the study is structured as follows. Section 4.2 presents the previous studies to give an overview of DL models against COVID-19. Moreover, this section shows the improvement in the accuracy as a contribution of the study by comparing a study that uses the same dataset. Section 4.3 briefly explains the DL methods and classification methods as classifiers. Section 4.4 presents the details of the case study. Section 4.5 presents some results and discussion. Finally, Section 4.6 concludes the study and gives some future directions.

4.2 LITERATURE REVIEW

Many studies agree that chest Computed Tomography (CT) and chest X-rays can be used in diagnosing COVID-19 patients (Abbas, Abdelsamea & Gaber, 2020; Apostolopoulos & Mpesiana, 2020; Asif et al., 2020; Chowdhury et al., 2020; Ko et al., 2020; Yoo et al., 2020). These images offer medical experts a more straightforward, faster, and reliable way to diagnose with the help of the wide availability of imaging machines in hospitals. CT and X-ray images enable us to classify the severity of COVID-19 by not just classifying COVID-19 positive or COVID-19 negative like laboratory test results. Besides, X-rays or CT can be performed without an increased risk of aerosolizing the pathogen, unlike laboratory tests (Hall et al., 2020).

Applying pre-trained models by transfer learning has been very common recently. Some authors proposed a COVID-19 classification model preferring just a pre-trained model as a baseline such as ResNet (Pathak et al., 2020), VGG16 (Civit-Masot et al., 2020; Panwar et al., 2020), Xception (Khan et al., 2020), Inception v3 (Asif et al., 2020). On the other hand, multiple pre-trained models were preferred in some studies. Ko et al. (2020) developed a FCONet, which is based on VGG16, ResNet-50, Inception v3, and Xception; Apostolopoulos and Mpesiana (2020) used VGG19, MobileNet-v2, Inception, Xception, and InceptionResNetv2. Hybrid models are also preferred in the literature. Hasan et al. (2020) extracted featured from images using Q-deformed entropy algorithm. The extracted features are classified with a long short-term memory (LSTM) neural network classifier, SVM, kNN, and logistic regression.

Some related studies are summarized in Table 4.1. The Test column refers to the accuracy score unless otherwise stated. The Data Size column gives the total number of images used in the research. The numbers of sub-groups given in the parenthesis are COVID-19, normal and other (such as pneumonia, lung cancer, SARS), respectively. Pre-trained models include VGG19, Inception, Inception v3, ResNetv2, ResNet-18, ResNet-101, MobileNetv2, SqueezeNet, ChezNet, and DenseNet-201. Neural Networks (NNs) are the most popular classification method. It is followed by SVM and DT. LSTM, kNN and logistic regression methods are other classifiers applied in COVID-19 diagnosis studies. The number of images in datasets varies in the studies. Due to the limited number of data, open-source datasets are used mainly by collecting data from platforms such as Kaggle and GitHub. Some authors comprised of dataset from multiple resources (Chowdhury et al., 2020; Yoo et al., 2020).

The multi-class classification accuracy of the studies, which use same dataset as our dataset is compared. Altan and Karasu (2020) used exactly same dataset as us. On the other hand, Asif et al. (2020) and Chowdhury et al. (2020) expand the same dataset

TABLE 4.1

Summary of Related Works

Study	Type	Model or Method	Classifier	Test (%)	Data Size
Hasan et al. (2020)	CT	Novel CNN	Several	99.68	321
Pathak et al. (2020)	CT	ResNet	NN	93	852
Barstugan et al. (2020)	CT	Several algorithms	SVM	99.68	150
Zheng et al. (2020)	CT	Novel CNN	NN	90.1	630
Wang et al. (2020a)	CT	COVID-19Net	NN	AUC:88	5,372
Ko et al. (2020)	CT	FCONet	NN	99.87	3,993
Civit-Masot et al. (2020)	X-Ray	VGG16	NN	86	396
Abbas et al. (2020)	X-Ray	DeTraC, ResNet	NN	95.12	196
Altan and Karasu (2020)	X-Ray	EfficientNet	NN	99.69	2,905
Khan et al. (2020)	X-Ray	Xception	NN	89.6	1,300
Panwar et al. (2020)	X-Ray	VGG16	NN	88.10	284
Yoo et al. (2020)	X-Ray	ResNet	DT	95	2,802
Ozturk et al. (2020)	X-Ray	DarkNet-19	NN	87.02	1,250
Tuncer et al. (2020)	X-Ray	LBP, ReliefF	Several	100.0	321
Apostolopoulos and Mpesiana (2020)	X-Ray	5 pre-trained models	NN	94.72	1,428
Wang and Wong (2020)	X-Ray	Novel CNN	NN	93.3	13,975
Sethy and Behera (2020)	X-Ray	13 pre-trained models	SVM	95.33	381
Narin et al. (2020)	X-Ray	Pre-trained models	NN	98	100
Hemdan et al. (2020)	X-Ray	7 pre-trained models	NN	90	50
Asif et al. (2020)	X-Ray	Inception v3	NN	93	3,550
Chowdhury et al. (2020)	X-Ray	8 pre-trained models	NN	97.94	3,487

with additional images. The proposed model gives accuracy of 100% with nine different hybrid models by extracting features from the images with pre-trained models and classifying them with various algorithms. Table 4.2 presents comparison results.

Studies in the literature attempt to develop hybrid methods to achieve high accuracy rates using these restricted datasets. Since the pandemic has spread rapidly, even the smallest improvements in the disease's diagnosis seem to be important in protecting people's health and managing the disease. With this motivation, pre-trained deep learning models, which provide beneficial results, especially in the feature extraction stage, were trained using a dataset consisting of 2,905 chest X-ray images, which include three classes: COVID-19, healthy, and viral pneumonia. After extracting features, four classification algorithms were run, which are Decision Tree (DT), Support Vector Machine (SVM), k-Nearest Neighbor (kNN), and the Self-Organizing Fuzzy Logic Classifier (SOFLC).

TABLE 4.2
Comparison Results

Study	Dataset	Models	3-class Accuracy (%)
Proposed model	Database (2020)	Hybrid models	100
Altan and Karasu (2020)	Database (2020)	EfficientNet-B0	99.69
Chowdhury et al. (2020)	Database (2020))+ additional images	DenseNet201	97.94
Asif et al. (2020)	Database (2020))+ additional images	Inception v3	93

4.3 METHODOLOGY

CNNs are specialized type of neural networks, have multiple layers, and can be applied to different number of dimensional data. CNNs are applied to extract features and offer a great support in the advancement in medical applications (Goodfellow, Bengio & Courville, 2016; Panwar et al., 2020). Figure 4.1 gives the CNN-based proposed model. Chest X-ray images are pre-processed at first. Then these images are fed into pre- trained CNNs for extracting features in the images. Lastly, extracted features are classified with various algorithms. Details of the proposed model and instructions about algorithms will be given in methods and case study sections.

A general approach in DL for a small number of image datasets is to use a pre-trained network. Transfer learning transfers knowledge mined in a pre-trained network to solve a different but related task, including new datasets that are smaller for training a CNN model from scratch. There are two commonly used ways to benefit from a pre-trained CNN: feature extraction and fine-tuning. Feature extraction with transfer learning refers to using the convolutional base of a pre-trained network with a new classifier and running the new dataset through it. The pre-trained model preserves its initial architecture and learned weights. The convolutional base is reusable because it learns representations which are generic for various tasks. In fine- tuning, some specific modifications are made to the pre-trained model. New trainable layers are added to the model while some of the knowledge mined from the previous task is preserved. These models require a large number of resources such as data and

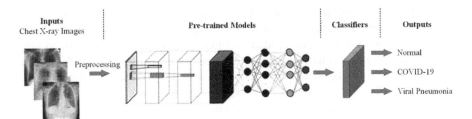

FIGURE 4.1 The proposed model.

computation power. In medical studies, both usages of pre-trained CNN are quite common (Apostolopoulos & Mpesiana, 2020; François, 2017).

4.3.1 DEEP LEARNING MODELS

In this study, the feature extraction is carried out utilizing 14 well-known pre-trained CNN models: AlexNet, DarkNet-53, DenseNet-201, GoogLeNet, Inception-Resnet-V2, Inception v3, MobileNet-V2, NASNet-Mobile, ResNet-18, ResNet-50, ResNet-101, ShuffleNet, SqueezeNet, and Xception. Brief introductory information of models is explained as below.

AlexNet is an 8-layer feedforward CNN model and competed in the ImageNet LSVRC-2010 contest. It has five convolution layers and ends with three fully connected layers. Some of the convolution layers in the model are followed by the max pooling layer (Krizhevsky, Sutskever & Hinton, 2012).

SqueezeNet was proposed to satisfy a given accuracy level with a small CNN architecture. It performs as well as AlexNet on ImageNet with fewer parameters. It also uses some model compression techniques which compress SqueezeNet 510 times smaller than AlexNet. SqueezeNet starts with convolution layer and is followed by fire modules. A fire module consists of a squeeze convolution layer. This layer has only a 1×1 filter, feeding into an expand layer that has a mixture of 1×1 and 3×3 convolution filters. Lastly, SqueezeNet ends with a final convolution layer (Iandola et al., 2016).

DarkNet-53 is a hybrid approach between the network used in YOLO-v2, DarkNet-19, and novel residual network stuff. It has 53 convolutional layers, and applies sequenced 1×1 and 3×3 convolutional layers. The network has some shortcut connections (Redmon & Farhadi, 2018).

In DenseNet, each layer is connected to every other layer in a feed-forward way. A traditional convolution network including L layers has L connections. On the other hand, a DenseNet network has L × (L+1)/2 connections. DenseNet architecture offers some advantages: it is beneficial for the vanishing gradient problem, enable feature reuse, reinforce feature propagation, and decrease the number of parameters considerably (Huang et al., 2017).

GoogLeNet is a 22 layer-deep network, begins with convolution layers, inception blocks follow them, and ends with a fully-connected layer. Inception blocks use filter sizes 1×1, 3×3 and 5×5. These modules enable to enhance the depth and width of the network while retaining the computational budget constant (Szegedy et al., 2015). GoogLeNet is named Inception v1. Various Inception architectures are
developed later. Inception v2 used batch normalization. Inception v3 uses additional factorized convolutions (Szegedy et al., 2016b).

ResNet network is based on residual learning framework. Residual networks are easy to optimize and achieve accuracy from rather increased depth. All types of ResNet-18, ResNet-50, ResNet-101, etc. are versions of ResNet. Their residual blocks scheme or blocks numbers differ (He et al., 2016).

Inception-ResNet combines both the Inception and Residual modules. Inception v3 does not use residual connections while Inception-ResNet-v1 and Inception-ResNet-v2 use residual connections instead of filter concatenation. Inception-ResNet-v1 is a hybrid Inception version and it has a comparable computational cost to Inception v3. On the other hand, Inception-ResNet-v2 is a costlier but increased recognition performance (Szegedy et al., 2016a).

MobileNet-v2 network puts importance of inverted residual and linear bottleneck blocks. The main part is a bottleneck depth separable convolution with residuals. Linear bottleneck helps the cases where non-linearity demolishes information in low dimensional space. The network includes the initial fully convolution layer with 32 filters, followed by residual bottleneck layers (Sandler et al., 2018).

NASNet is inspired by the Neural Architecture Search framework (Zoph & Le, 2016). NASNet uses a reinforcement learning search approach to optimize architecture configurations (Zoph et al., 2018).

ShuffleNet is developed especially for the devices with low computing power. The network uses two novel operations: pointwise group convolution and channel shuffle. These operations decrease computation cost while sustaining accuracy. Specifically, group convolution decreases computation complexity of 1×1 convolutions. Shuffle operation is used to support the information owing across feature channel. ShuffleNet lets more feature map channels that enables to encode more information (Zhang et al., 2018).

Xception, means "*Extreme Inception*", inspired by Inception where Inception blocks have been changed with depthwise separable convolution layers. These depthwise separable convolution layers are stacked linearly with residual connections. The network has 36 convolutional layers which forms the feature extraction base (Chollet, 2017).

4.3.2   CLASSIFIERS

Extracted features with the prementioned fourteen pre-trained CNNs are classified with Decision Trees (DTs), Support Vector Machine (SVM), k-Nearest Neighbor (kNN), and Self-Organizing Fuzzy Logic Classifier (SOFLC). A detailed explanation of computations is beyond the scope of this study. The algorithms are explained briefly.

Decision Trees (DTs) are a tree-based and supervised machine learning algorithm. The aim of DTs is predicting the target value by learning the decision rules concluded from the features of data. DTs can overcome both categorical and numerical data. DTs apply a white box model. When a given situation is observable in a model, the condition's explanation is described simply by Boolean logic. (Scikit-learn, 2020b).

Support Vector Machine (SVM) are effective algorithms in performing linear and non-linear classification problems. In its decision function, it uses a subset of training points, also called support vectors. The support vectors provide memory efficiency in the learning phase of SVMs. In non-linear problems, SVMs use kernel trick. SVMs use hyperplanes in separating two classes. Optimal separating is achieved when hyperplane maximizes the distance to the closest point from either two classes (Gunn et al., 1998; Yazici et al., 2020).

k-Nearest Neighbor (kNN) uses neighbor methods and is a non-parametric method. The input of kNNs consists of k neighbors from the feature space. The outputs depend on problem type: regression or classification problem. In classification problems, the output of kNN gives a class membership. The main idea behind kNN is to find a predefined number of training samples closest in distance to the new point, and then prediction of the label is made from these. The number of samples can be a constant, specified by the user, or variate based on the local density of points. Any metric can be used in distance calculation; standard Euclidean distance is the most preferred choice (Scikit-learn, 2020a).

The Self-Organizing Fuzzy Logic Classifier (SOFLC) is a self-organizing non-parametric fuzzy rule-based classifier. The first SOFLC was proposed by (Procyk & Mamdani, 1979). It can adjust its basic fuzzy logic rule-base subjected to its experience and environment. The SOFLC implements two tasks together. It observes the environment while allowing the proper control action and uses the results of actions to develop them even more. SOFLC architecture includes two essential parts. The first part is a simple fuzzy controller. The following one builds upon the self-organizing mechanism, which functioning as a monitor and an evaluator of the low-level controller performance (Lu & Mahfouf, 2010).

4.4 CASE STUDY

4.4.1 Dataset

The publicly available dataset used in this study includes 219 COVID-19, 1,345 viral pneumonia, and 1,341 normal chest x-ray images (Database, 2020). Whereas the number of images for the healthy and viral pneumonia is near to each other, COVID-19 images are meager. In this study, the dataset images are scaled to the dimensions that can be used in the training and test stages. Then, the scaled data were used to train the models. The models and their features are given in Table 4.3.

TABLE 4.3

Deep Learning Models and Properties

Pre-trained Model	Depth	Parameters (Millions)	Input Image Size
AlexNet	8	61	227 × 227 × 3
DarkNet-53	53	41	256 × 256 × 3
DenseNet-201	201	20	224 × 224 × 3
GoogLeNet	22	7	224 × 224 × 3
Inception-ResNet-v2	164	55.9	229 × 229 × 3
Inception-v3	48	23.9	229 × 229 × 3
MobileNet-v2	53	3.5	224 × 224 × 3
NASNet-Mobile[a]	-	5.3	224 × 224 × 3
ResNet-18	18	11.7	224 × 224 × 3
ResNet-50	50	25.6	224 × 224 × 3
ResNet-101	101	44.6	224 × 224 × 3
ShuffleNet	50	1.4	224 × 224 × 3
SqueezeNet	18	1.24	227 × 227 × 7
Xception	71	22.9	229 × 229 × 3

[a] NASNet-Mobile model does not consist of a linear sequence of module

4.4.2 PRELIMINARIES

The data set used in the study contains 219, 1,341, and 1,345 chest X-ray images of COVID-19, healthy and viral pneumonia diseases, respectively. The depth parameter represents the largest number of sequential convolutional or fully connected layers on a path from the input layer to the output layer. One of the classification achievements, accuracy as an illustrative example obtained by re-training the models given in Table 4.3 with the data set discussed are given in Figure 4.2. The accuracy of pre-trained models varies from 92.49% to 96.56%. These values indicate that such critical disease data cannot be fully classified.

The features are extracted from the trained deep learning model and classified through the DT, SVM, kNN, and SOFLC classifiers to increase the accuracy values and make a more accurate classification. The TotalBoost method was chosen as the ensemble aggregation method, and the 100 learners were selected for the DT classifier. For SVM, a linear kernel is preferred. In the kNN and SOFLC model, the Euclidean distance function is used. Granularity value was determined as 12 in SOFCL. Figure 4.3 represents the increased accuracy scores.

Within the study's scope, results were obtained using 10-fold cross-validation to avoid overfitting. The training data were divided into pieces in the cross-validation process and used for validation to examine whether the classifiers are dependent on the data. Seven main performance metrics, accuracy, sensitivity, specificity, precision, false positive rate, F1-score Matthew's correlation coefficient (MCC), are measured in this study. Accuracy, which defines systematic errors and measures the statistics of bias (Chatrati et al., 2020). F1 score is the harmonic mean of the recall and precision, which are other two performance metrics (Doğan, 2019). Accuracy and F1 score are popularly measured metrics in classification problems. However,

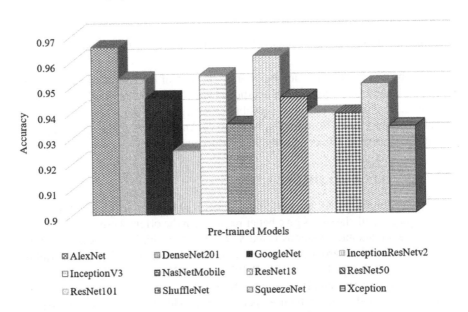

FIGURE 4.2   Classification accuracy of the pre-trained models.

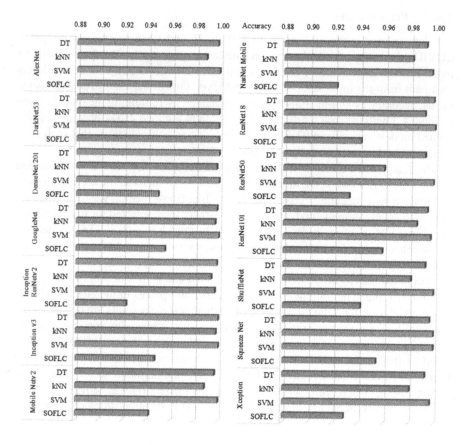

FIGURE 4.3   Classification accuracy of the pre-trained models with classifiers.

they can be less informative due to ignoring the size of the confusion matrix's four classes. The Matthews correlation coefficient (MCC) considers true/false positives/negatives. It is mainly seen as an equilibrated metric that can be also calculated for unbalanced class sizes (Boughorbel, Jarray & El-Anbari, 2017). The remaining measurement types were also considered to see different results.

4.5   RESULTS AND DISCUSSIONS

The calculation results for the performance metrics are given in Table 4.4. It is seen that, in general, all classifiers give better results than the original pre-trained models. One of the remarkable situations is the SVM classifier's high success in classifying the features obtained from AlexNet, GoogLeNet, MobileNet-v2, NasNet Mobile, ResNet-18, ResNet-50, ResNet-101, ShuffleNet, and Xception models. DT classifier works with DarkNet53 and Inception ResNetv2 models. Besides, DenseNet 201 and Inception v3 models have highest performance with SVM and DT classifiers. SqueezeNet is the only model that has the highest performance with kNN. Another

TABLE 4.4

Performance Metrics for Classifiers

Base Model	Classifier	Acc. (%)	Sens. (%)	Spec. (%)	Prec. (%)	FPR (%)	F1 (%)	MCC (%)
AlexNet	DT	99.85	99.71	99.91	99.89	0.09	99.80	99.72
	kNN	98.92	97.40	99.33	99.23	0.67	98.27	97.69
	SVM	100.00	100.00	100.00	100.00	0.00	100.00	100.00
	SOFLC	95.87	93.64	97.47	96.59	2.53	95.01	92.65
DarkNet53	DT	100.00	100.00	100.00	100.00	0.00	100.00	100.00
	kNN	99.95	99.97	99.98	99.78	0.02	99.87	99.85
	SVM	99.95	99.97	99.98	99.78	0.02	99.87	99.85
	SOFLC	99.95	99.97	99.98	99.78	0.02	99.87	99.85
DenseNet 201	DT	100.00	100.00	100.00	100.00	0.00	100.00	100.00
	kNN	99.80	99.86	99.89	99.68	0.11	99.77	99.65
	SVM	100.00	100.00	100.00	100.00	0.00	100.00	100.00
	SOFLC	94.95	92.56	96.99	94.56	3.01	93.49	90.61
GoogleNet	DT	99.85	99.89	99.91	99.89	0.09	99.89	99.80
	kNN	99.71	99.79	99.82	99.79	0.18	99.79	99.60
	SVM	100.00	100.00	100.00	100.00	0.00	100.00	100.00
	SOFLC	95.53	92.13	97.41	94.01	2.59	93.02	90.53
Inception ResNetv2	DT	99.85	99.71	99.92	99.71	0.08	99.71	99.63
	kNN	99.41	99.03	99.72	98.34	0.28	98.68	98.38
	SVM	99.71	99.42	99.87	99.07	0.13	99.24	99.10
	SOFLC	92.32	88.12	95.32	93.04	4.68	90.25	85.97
Inception v3	DT	100.00	100.00	100.00	100.00	0.00	100.00	100.00
	kNN	99.80	99.86	99.90	99.50	0.10	99.68	99.57
	SVM	100.00	100.00	100.00	100.00	0.00	100.00	100.00
	SOFLC	94.73	93.66	96.82	94.91	3.18	94.25	91.15
Mobile Netv2	DT	99.71	99.42	99.84	99.42	0.16	99.42	99.27
	kNN	98.87	99.19	99.49	96.74	0.51	97.89	97.33
	SVM	100.00	100.00	100.00	100.00	0.00	100.00	100.00
	SOFLC	94.27	92.48	96.56	94.12	3.44	93.25	89.93
NasNet Mobile	DT	99.31	99.14	99.60	99.14	0.40	99.14	98.74
	kNN	98.23	98.72	99.12	95.99	0.88	97.27	96.34
	SVM	99.75	99.64	99.87	99.46	0.13	99.55	99.42
	SOFLC	92.20	90.99	95.46	90.35	4.54	90.64	86.12
ResNet18	DT	99.90	99.93	99.94	99.93	0.06	99.93	99.87
	kNN	99.21	99.43	99.54	99.08	0.46	99.25	98.78
	SVM	100.00	100.00	100.00	100.00	0.00	100.00	100.00
	SOFLC	94.15	91.56	96.49	94.00	3.51	92.67	89.34
ResNet50	DT	99.26	99.29	99.57	99.11	0.43	99.20	98.76
	kNN	96.02	97.13	97.97	92.30	2.04	94.36	92.40
	SVM	99.90	99.93	99.95	99.75	0.05	99.84	99.78
	SOFLC	93.23	90.05	95.98	92.46	4.02	91.12	87.34
ResNet101	DT	99.46	99.06	99.69	99.24	0.31	99.15	98.85
	kNN	98.62	99.01	99.20	98.30	0.80	98.65	97.82
	SVM	99.71	99.61	99.82	99.79	0.18	99.70	99.52
	SOFLC	95.87	94.49	97.53	95.70	2.47	95.07	92.66
ShuffleNet	DT	99.31	98.96	99.59	99.32	0.41	99.14	98.74
	kNN	98.18	98.32	98.94	97.81	1.06	98.06	96.99
	SVM	99.95	99.78	99.97	99.97	0.03	99.87	99.85
	SOFLC	94.15	90.71	96.40	95.33	3.60	92.77	89.47

(Continued)

TABLE 4.4 (Continued)

Base Model	Classifier	Acc. (%)	Sens. (%)	Spec. (%)	Prec. (%)	FPR (%)	F1 (%)	MCC (%)
	DT	99.66	99.39	99.79	99.75	0.21	99.57	99.37
Squeeze	kNN	99.95	99.97	99.97	99.97	0.03	99.97	99.93
Net	SVM	99.95	99.97	99.97	99.97	0.03	99.97	99.93
	SOFLC	95.41	92.47	97.36	93.61	2.64	93.01	90.45
	DT	99.31	99.32	99.59	99.32	0.41	99.32	98.91
	kNN	98.08	98.62	98.95	96.79	1.05	97.65	96.58
Xception	SVM	99.71	99.79	99.82	99.79	0.18	99.79	99.60
	SOFLC	92.89	91.49	95.71	93.14	4.29	92.25	88.09

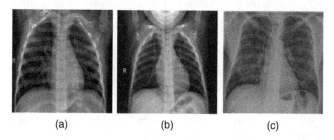

(a)   (b)   (c)

FIGURE 4.4   Chest X-rays for (a) COVID-19, (b) normal and (c) viral pneumonia classes.

interesting case is that the SOFLC classifier's success is near to the pre-trained models at some points but higher than GoogLeNet, ResNet-101, ShuffleNet and SqueezeNet models. The table's values show that pre-trained models produce successful results in feature extraction, while traditional classifiers produce successful results in classification.

One randomly selected image from COVID-19, normal and viral pneumonia classes was used to test the trained model (Figure 4.4). Images were classified using the retrained AlexNet model. The probabilities of belonging to three classes are given in the order of COVID-19, normal, viral pneumonia, which obtained as (0, 0.96, 0.04), (1, 0, 0) and (0.01, 0, 0.99) for Figure 4.4a, b (0.96, 0.04, 0) and Figure 4.4c, respectively. In the classification, areas where the model is concentrated and effective in the classification are colored.

Figure 4.5 presents the densities of the areas corresponding to the X-ray images given in Figure 4.4. The X-ray chest image given in Figure 4.4b was classified with AlexNet, and the features obtained from the deep learning model were placed on the top of the image in Figure 4.5a. The same procedures were repeated for images belonging to the normal and viral pneumonia classes, and the images in Figure 4.5b and c were obtained, respectively. While the trained model handles the test data for classification, it concentrates on specific parts of the image to determine which class it belongs. The colors scaled from red to blue in the images represent the effect of the corresponding areas on the classification. The red areas in Figure 4.5 show the height of the matching of the features in the relevant region, while the blue areas show that the feature match is low.

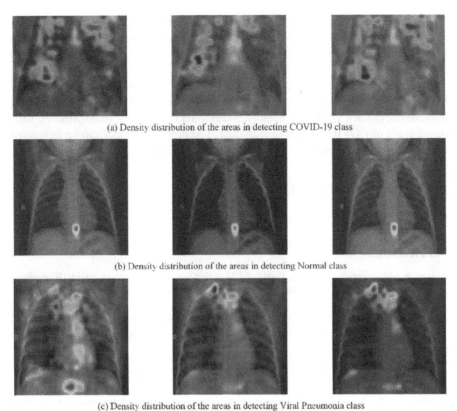

(a) Density distribution of the areas in detecting COVID-19 class

(b) Density distribution of the areas in detecting Normal class

(c) Density distribution of the areas in detecting Viral Pneumonia class

FIGURE 4.5 Density distribution of areas in detecting classes.

4.6 CONCLUSION AND FUTURE DIRECTIONS

COVID-19 caused more than 20 million people to get the disease and more than 800 thousand people to die due to its high spreading rate. Therefore, academic studies for the detection of COVID-19 are of great importance. In this study, the classification of COVID-19, normal, and viral pneumonia diseases was made using a dataset consisting of chest X-ray images obtained by various researchers. In the study, 14 different deep learning models were trained, and the properties obtained with these deep learning models were taken by using the transfer learning method. Four different classifiers were used in the classification of these data. 10-fold cross-validation was used to prevent overfitting and data-dependency in the training of classifiers. In the training and test phases of the proposed hybrid networks, 70% and 30% of the data in the data set were used. The results obtained in the study show that higher success can be achieved by using deep learning models and classifier models as a hybrid. Higher performance of the hybrid method over single deep learning models is shown in comparison results in Table 4.2. The proposed model gives accuracy of 100% with nine different hybrid models, such as ResNet18 with SVM, MobileNet-v2 with

SVM, Inception v3 with DT, etc. The rest of hybrid models also present satisfactory performance. SVM is found the most successful classifier among the four classifiers. It is thought that the model obtained in this study will help to detect COVID-19 disease.

This study aims to expand dataset for future studies not only by using open data, but also trying to collect data from hospitals in Turkey. Classification is made taking into consideration COVID-19, healthy, pneumonia images. For future studies, various degrees of COVID-19 can be classified. The proposed model can be integrated into smart devices because of its accurate and cost-effective performance.

DISCLOSURE STATEMENT

Authors declare that they have no conflict of interest.

REFERENCES

Abbas, A., Abdelsamea, M.M. & Gaber, M.M. (2020). Classification of covid-19 in cjhest x-ray images using detrac deep convolutional neural network. *Applied Intelligence*, 51, 854–864.

Altan, A. & Karasu, S. (2020). Recognition of Covid-19 disease from x-ray images by hybrid model consisting of 2D curvelet transform, chaotic salp swarm algorithm and deep learning technique. *Chaos, Solitons & Fractals*, 140, 110071.

Apostolopoulos, I.D. & Mpesiana, T.A. (2020). Covid-19: Automatic detection from x-ray images utilizing transfer learning with convolutional neural networks. *Physical and Engineering Sciences in Medicine*, 43, 635–640.

Ardakani, A.A., Kanafi, A.R., Acharya, U.R., Khadem, N. & Mohammadi, A. (2020). Application of deep learning technique to manage covid-19 in routine clinical practice using ct images: Results of 10 convolutional neural networks. *Computers in Biology and Medicine*, 121, 103795.

Asif, S., Wenhui, Y., Jin, H., Tao, Y. & Jinhai, S. (2020). Classification of covid-19 from chest x-ray images using deep convolutional neural networks. medRxiv: 2020.05.01.20088211.

Barstugan, M., Ozkaya, U., & Ozturk, S. (2020). Coronavirus (covid-19) classification using ct images by machine learning methods. arXiv preprint arXiv:2003.09424.

Bernheim, A., Mei, X., Huang, M., Yang, Y., Fayad, Z.A., Zhang, N., … et al. (2020). Chest CT findings in coronavirus disease-19 (covid-19): Relationship to duration of infection. *Radiology*, 295, 200463.

Boughorbel, S., Jarray, F. & El-Anbari, M. (2017). Optimal classifier for imbalanced data using matthews correlation coefficient metric. *PloS one*, 12 (6), e0177678.

Butt, C., Gill, J., Chun, D. & Babu, B.A. (2020). Deep learning system to screen coronavirus disease 2019 pneumonia. *Applied Intelligence*, 1.

Chatrati, S.P., Hossain, G., Goyal, A., Bhan, A., Bhattacharya, S., Gaurav, D. & Tiwari, S.M. (2020). Smart home health monitoring system for predicting type 2 diabetes and hypertension. *Journal of King Saud University-Computer and Information Sciences*.

Chollet, F. (2017). *Xception: Deep learning with depthwise separable convolutions*. In *Proceedings of the IEEE conference on computer vision and pattern recognition* (pp. 1251–1258).

Chowdhury, M.E., Rahman, T., Khandakar, A., Mazhar, R., Kadir, M.A., Mahbub, Z.B., … et al. (2020). Can ai help in screening viral and covid-19 pneumonia? *IEEE Access*, 8, 132665–132676.

Civit-Masot, J., Luna-Perejon, F., Dominguez Morales, M. & Civit, A. (2020). Deep learning system for covid-19 diagnosis aid using x-ray pulmonary images. *Applied Sciences*, 10 (13), 4640.

Corman, V.M., Landt, O., Kaiser, M., Molenkamp, R., Meijer, A., Chu, D.K., … et al. (2020). Detection of 2019 novel coronavirus (2019-nCoV) by real-time RT-PCR. *Euro-Surveillance*, 25 (3), 2000045.

Coronaviruses. (2020). https://www.niaid.nih.gov/diseases-conditions/coronaviruses. (Accessed: 2020-05-30)

Database, C-.R. (2020). https://www.kaggle.com/tawsifurrahman/covid19-radiography-database? . (Accessed: 2020-08-01)

Doğan, O. (2019). Data linkage methods for big data management in industry 4.0. In Oner, S.C. & Yuregir, O.H. (Eds.) *Optimizing big data management and industrial systems with intelligent techniques* (pp. 108–127). IGI Global.

Fan, Y., Zhao, K., Shi, Z.-L. & Zhou, P. (2019). Bat coronaviruses in china. *Viruses*, 11 (3), 210.

Fang, Y., Zhang, H., Xie, J., Lin, M., Ying, L., Pang, P. & Ji, W. (2020). Sensitivity of chest CT for covid-19: comparison to RT-PCR. *Radiology*, 200432.

François, C. (2017). *Deep learning with python*. Manning Publications Company.

Goodfellow, I., Bengio, Y. & Courville, A. (2016). *Deep learning*. MIT press.

Gozes, O., Frid-Adar, M., Greenspan, H., Browning, P.D., Zhang, H., Ji, W., …Siegel, E. (2020). Rapid AI development cycle for the coronavirus (covid-19) pandemic: Initial results for automated detection & patient monitoring using deep learning ct image analysis. arXiv preprint arXiv:2003.05037.

Gunn, S.R. et al. (1998). Support vector machines for classification and regression. *ISIS technical report*, 14 (1), 5–16.

Hall, L.O., Paul, R., Goldgof, D.B., & Goldgof, G.M. (2020). Finding covid-19 from chest x-rays using deep learning on a small dataset. arXiv preprint arXiv:2004.02060.

Hasan, A.M., AL-Jawad, M.M., Jalab, H.A., Shaiba, H., Ibrahim, R.W. & AL-Shamasneh, A.R. (2020). Classification of covid-19 coronavirus, pneumonia and healthy lungs in ct scans using q-deformed entropy and deep learning features. *Entropy*, 22 (5), 517.

He, K., Zhang, X., Ren, S. & Sun, J. (2016). *Deep residual learning for image recognition*. In *Proceedings of the IEEE conference on computer vision and pattern recognition* (pp. 770–778).

Hemdan, E.E.-D., Shouman, M.A. & Karar, M.E. (2020). Covidx-net: A framework of deep learning classifiers to diagnose covid-19 in x-ray images. arXiv preprint arXiv:2003.11055.

Huang, G., Liu, Z., Van Der Maaten, L. & Weinberger, K.Q. (2017). *Densely connected convolutional networks*. In *Proceedings of the ieee conference on computer vision and pattern recognition* (pp. 4700–4708).

Iandola, F.N., Han, S., Moskewicz, M.W., Ashraf, K., Dally, W.J. & Keutzer, K. (2016). Squeezenet: Alexnet-level accuracy with 50x fewer parameters and< 0.5 mb model size. arXiv preprint arXiv:1602.07360.

Iqbal, H., Romero-Castillo, K. D., Bilal, M. & Parra-Saldivar, R. (2020). The emergence of novel-coronavirus and its replication cycle: an overview. *Journal of Pure and Applied Microbiology*, 14 (1), 13–16.

Khan, A.I., Shah, J.L. & Bhat, M.M. (2020). Coronet: A deep neural network for detection and diagnosis of covid-19 from chest x-ray images. *Computer Methods and Programs in Biomedicine*, 196, 105581.

Ko, H., Chung, H., Kang, W.S., Kim, K.W., Shin, Y., Kang, S.J., … et al. (2020). Covid-19 pneumonia diagnosis using a simple 2D deep learning framework with a single chest CT image: model development and validation. *Journal of Medical Internet Research*, 22 (6), e19569.

Krizhevsky, A., Sutskever, I. & Hinton, G. E. (2012). Imagenet classification with deep convolutional neural networks. In Jordan, M.I., LeCun, Y. & Saran, A.S. (Eds.). *Advances in neural information processing systems* (pp. 1097–1105).

Lan, L., Xu, D., Ye, G., Xia, C., Wang, S., Li, Y. & Xu, H. (2020). Positive rt-pcr test results in patients recovered from covid-19. *Jama*, 323 (15), 1502–1503.

Lu, Q. & Mahfouf, M. (2010). A model-free self-organizing fuzzy logic control system using a dynamic performance index table. *Transactions of the Institute of Measurement and Control*, 32 (1), 51–72.

Mahase, E. (2020). *Coronavirus: Covid-19 has killed more people than sars and mers combined, despite lower case fatality rate.* British Medical Journal Publishing Group.

Narin, A., Kaya, C. & Pamuk, Z. (2020). Automatic detection of coronavirus disease (covid-19) using x-ray images and deep convolutional neural networks. arXiv preprint arXiv:2003.10849.

Ozturk, T., Talo, M., Yildirim, E.A., Baloglu, U.B., Yildirim, O. & Acharya, U.R. (2020). Automated detection of covid-19 cases using deep neural networks with x-ray images. *Computers in Biology and Medicine*, 121, 103792.

Panwar, H., Gupta, P., Siddiqui, M.K., Morales-Menendez, R. & Singh, V. (2020). Application of deep learning for fast detection of covid-19 in x-rays using ncovnet. *Chaos, Solitons & Fractals*, 109944.

Pathak, Y., Shukla, P.K., Tiwari, A., Stalin, S., Singh, S. & Shukla, P.K. (2020). Deep transfer learning based classification model for covid-19 disease. *IRBM*.

Procyk, T.J. & Mamdani, E.H. (1979). A linguistic self-organizing process controller. *Automatica*, 15 (1), 15–30.

Redmon, J. & Farhadi, A. (2018). Yolov3: An incremental improvement. arXiv preprint arXiv:1804.02767.

Sandler, M., Howard, A., Zhu, M., Zhmoginov, A. & Chen, L.-C. (2018). *Mobilenetv2: Inverted residuals and linear bottlenecks.* In *Proceedings of the IEEE conference on computer vision and pattern recognition* (pp. 4510–4520).

Scikit-learn. (2020a). https://scikit-learn.org/stable/modules/neighbors.html#unsupervised-neighbors. (Accessed: 2020-08-08)

Scikit-learn. (2020b). https://scikit-learn.org/stable/modules/tree.html#tree. (Accessed: 2020-08-08)

Sethy, P.K. & Behera, S.K. (2020). Detection of coronavirus disease (covid-19) based on deep features. Preprints, 2020030300, 2020.

Siddiqui, M.K., Morales-Menendez, R., Gupta, P.K., Iqbal, H., Hussain, F., Khatoon, K. & Ahmad, S. (2020). Correlation between temperature and covid-19 (suspected, confirmed and death) cases based on machine learning analysis. *Journal of Pure and Applied Microbiology*, 14, 1017–1024.

Szegedy, C., Ioffe, S., Vanhoucke, V. & Alemi, A. (2016a). Inception-v4, inception-resnet and the impact of residual connections on learning. arXiv preprint arXiv:1602.07261.

Szegedy, C., Liu, W., Jia, Y., Sermanet, P., Reed, S., Anguelov, D., ... Rabinovich, A. (2015). *Going deeper with convolutions.* In *Proceedings of the IEEE conference on computer vision and pattern recognition* (pp. 1–9).

Szegedy, C., Vanhoucke, V., Ioffe, S., Shlens, J. & Wojna, Z. (2016b). *Rethinking the inception architecture for computer vision.* In *Proceedings of the ieee conference on computer vision and pattern recognition* (pp. 2818–2826).

Tuncer, T., Dogan, S. & Ozyurt, F. (2020). An automated residual exemplar local binary pattern and iterative relieff based corona detection method using lung x-ray image. *Chemometrics and Intelligent Laboratory Systems*, 203, 104054.

Wang, L. & Wong, A. (2020). Covid-net: A tailored deep convolutional neural net-work design for detection of covid-19 cases from chest x-ray images. *Scientific Reports*, 10, 19549.

Wang, S., Zha, Y., Li, W., Wu, Q., Li, X., Niu, M., ... et al. (2020a). A fully automatic deep learning system for covid-19 diagnostic and prognostic analysis. *European Respiratory Journal*.

Wang, W., Xu, Y., Gao, R., Lu, R., Han, K.,Wu, G. & Tan, W. (2020b). Detection of sars-cov-2 in different types of clinical specimens. *Jama*, 323 (18), 1843–1844.

WHO. (2020). *R&D Blueprint and COVID-19*, https://www.who.int/teams/blueprint/covid-19. (Accessed: 2020-07-09)

Xie, X., Zhong, Z., Zhao, W., Zheng, C., Wang, F. & Liu, J. (2020). Chest CT for typical 2019-ncov pneumonia: relationship to negative RT-PCR testing. *Radiology*, 296 (2), E41–E45

Yazici, I., Beyca, O.F., Gurcan, O.F., Zaim, H., Delen, D. & Zaim, S. (2020). A comparative analysis of machine learning techniques and fuzzy analytic hierarchy process to determine the tacit knowledge criteria. *Annals of Operations Research*, 1–24.

Yoo, S.H., Geng, H., Chiu, T. L., Yu, S.K., Cho, D.C., Heo, J., ... et al. (2020). Deep learning-based decision-tree classifier for covid-19 diagnosis from chest x-ray imaging. *Frontiers in Medicine*, 7, 427.

Zhang, X., Zhou, X., Lin, M. & Sun, J. (2018). *Shufflenet: An extremely efficient convolutional neural network for mobile devices*. In *Proceedings of the ieee conference on computer vision and pattern recognition* (pp. 6848–6856).

Zheng, C., Deng, X., Fu, Q., Zhou, Q., Feng, J., Ma, H., ...Wang, X. (2020). Deep learning-based detection for covid-19 from chest ct using weak label. medRxiv.

Zhu, H., Guo, Q., Li, M., Wang, C., Fang, Z., Wang, P., ...Xiao, Y. (2020). Host and infectivity prediction of wuhan 2019 novel coronavirus using deep learning algorithm. BioRxiv: 2020.01.21.914044.

Zoph, B. & Le, Q. V. (2016). Neural architecture search with reinforcement learning. arXiv preprint arXiv:1611.01578.

Zoph, B., Vasudevan, V., Shlens, J. & Le, Q. V. (2018). *Learning transferable architectures for scalable image recognition*. In *Proceedings of the ieee conference on computer vision and pattern recognition* (pp. 8697–8710).

Zu, Z.Y., Jiang, M.D., Xu, P.P., Chen, W., Ni, Q.Q., Lu, G.M. & Zhang, L.J. (2020). Coronavirus disease 2019 (covid-19): a perspective from china. *Radiology*, 296 (2), E15 E25.

5 Deep Learning in BioMedical Applications
Detection of Lung Disease with Convolutional Neural Networks

Emre Olmez
Yozgat Bozok University, Turkey

Orhan Er and Abdulkadir Hiziroglu
İzmir Bakırçay University, Turkey

CONTENTS

5.1 INTRODUCTION

In this study, an artificial intelligence application was developed for the detection of lung diseases, which are quite prevalent in Turkey and the World to provide an example for biomedical applications of deep learning methods.

DOI: 10.1201/9781003161233-5

Millions of people are diagnosed every year with a lung disease in the world. Asthma, chronic obstructive lung disease, pneumonia, tuberculosis (TB) and lung cancer are most important lung diseases. And these are very common illnesses in Turkey and the World (WHO, 2021). It is very valuable to develop a decision support system with an artificial intelligence method for the diagnosis of these diseases.

Although there are many different studies on this subject using ordinary artificial neural networks architecture or image-based computer-aided diagnosis, this study is valuable because there is no study in the literature on the classification of lung disease using convolutional neural networks using clinical data (Ashizawa et al., 1999; Khobragade et al., 2016; Hattikatti, 2017; Kido et al., 2018; Geng et al., 2019).

Kido et al. (2018) developed an image-based computer-aided detection (CADe) algorithm by use of regions with CNN features (R-CNN) for detection of lung abnormalities. They evaluated the performance of image-based CADe by use of R-CNN for various kinds of lung abnormalities such as lung nodules and diffuse lung diseases.

Geng et al., 2019 developed a 2.5D CNN model to detect multiple diseases in multiple organs in CT scans. In this study they investigated detection of four common diseases in the lungs, which are atelectasis, edema, pneumonia and nodule. They also implemented a train-validation-test process for each disease to evaluate the generalization of their model and again got comparable test results, 0.818 for atelectasis, 0.963 for edema, 0.878 for pneumonia and 0.784 for nodule.

Hattikatti presented a new system for interstitial lung disease detection. They designed CNN to classify 2D lung images. The proposed system gave promising results on dataset of 30 CT scan images using CNN which had 94% accuracy with high sensitivity. SVM classifier on the same dataset give the result of 86% accuracy (Hattikatti, 2017).

Ashizawa et al., 1999 developed a new method to distinguish between various interstitial lung diseases that uses an artificial neural network. This network is based on features extracted from chest radiographs and clinical parameters. The aim of their study was to evaluate the effect of the output from the artificial neural network on radiologists' diagnostic accuracy.

A paper of Khobragade et al. proposed lung segmentation; lung feature extraction and its classification using artificial neural network technique for the detection of lung diseases such as tuberculosis, lung cancer and pneumonia. They have used the simple image processing techniques like intensity-based method and discontinuity-based method to detect lung boundaries (Khobragade et al., 2016).

Some studies in the literature are given above to reveal the original approach of this study. As can be understood from these studies, no article trying to determine the classification problem of lung diseases using CNN structure with clinical data.

Lung diseases constitute a significant portion of chronic diseases. Respiratory diseases are the third most common cause of death in Turkey. Due to causes such as air pollution, smoking, lack of physical activity, disorganized life, and more recently, the damage caused by the COVID-19 virus, chronic lung diseases have become frequent. It is predicted that following the periods of little activity in limited areas made mandatory by the pandemic conditions, there will be an acceleration in the increases of lung symptoms with negative characteristics.

Most of the studies in recent years have focused on the diagnosis of chest graphs with artificial intelligence methods. In these studies, image processing and deep learning methods were used together. Contrary to these studies, our study aspired to create a decision support system for the diagnosis of the diseases in question. By using the data obtained from the examinations performed by the specialist physicians besides, it was sought to create an example regarding how biomedical data may be processed with deep learning algorithms.

The most common lung diseases were realized by using convolution neural network (CNN) which was found to be of high performance as a result of the literature review (Arena et al., 2003; Liang et al., 2016; Wimmer et al., 2016; Gao et al., 2018; Oh et al., 2018). In order to demonstrate the CNN success, the lung dataset was used with a feedback artificial neural network and the results were compared. For this purpose, a data set consisting of 357 subjects (patients), with 100 being in normal condition and 257 having various lung diseases, was obtained from a local city hospital and automatic disease detection was performed by using the Convolutional Neural Networks (CNN) method, which is one of the most frequently used deep learning methods and for comparison, a probabilistic artificial neural network (PNN) model was used (Speckt, 1990). All samples included in the data set have 38 characteristics determined by specialist physicians. The proposed system is given in Figure 5.1 and has been prepared to adapt to an infrastructure suitable for use in second and third level healthcare institutions and digital hospitals.

Since inputs to CNNs are made using images, the input data was transformed into a square matrix and applied to the CNN. First, a CNN architecture with a 6 × 6 square matrix input was designed, which allowed the selection of 36 of the 38 features of the subjects. The designed CNN was individually trained with all combinations of 36 of 38 features (703 in total), and the feature set with the highest success rate was selected for this model. Afterwards, a parameter selection was carried out using eight different models trained using this feature set. Lastly, the success of the study was demonstrated using the changes made to the network architecture.

In the organization of this chapter, first, the importance and reasoning behind this study re stated. Then, lung diseases are introduced in the second section. In the third section, the used CNN architecture and its training are described. The architecture and parameters determined for the application are explained in the fourth section, and in the final section, the results of the study are evaluated.

5.2 LUNG DISEASES

Lung diseases are associated with the diseases of the respiratory system. The respiratory system is a system consisting of lungs and respiratory tracts, through which oxygen, the main fuel for burning the food taken into the body, is procured and the resulting carbon dioxide is removed from the body.

The most common lung diseases (Khobragade et al., 2016; Sen et al., 2020) treated and followed up in lung disease clinics as outpatient cases, or inpatient cases when necessary, are as follows:

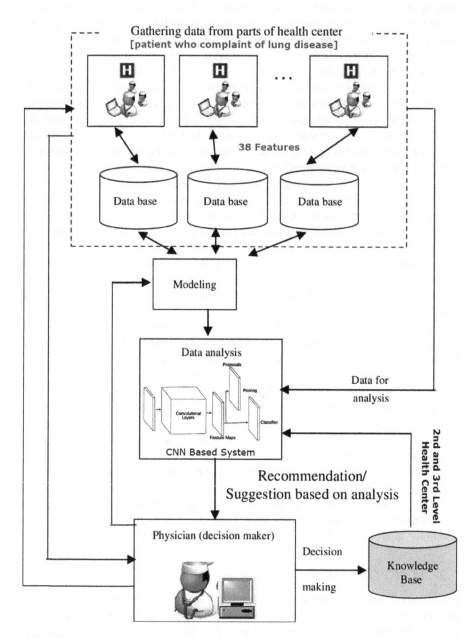

FIGURE 5.1 Deep learning method for the lung disease.

Asthma
Chronic obstructive lung disease (chronic bronchitis and emphysema-COPD)
Pneumonia
Tuberculosis (TB)
Lung Cancer

1. *Asthma* typically causes coughing, a whistling sound when breathing in and out, most significantly, shortness of breath. The characteristic feature of asthma complaints is that they are in the form of crises that occur sporadically. This is due to the highly sensitized bronchial mucous membrane, which is weakened by the returning inflammation (Er, 2009; Yılmaz, 2012). Asthma is also a chronic disease characterized by recurrent attacks of breathlessness and wheezing. During an asthma attack, the lining of the bronchial tubes swell, causing the airways to narrow, reducing the flow of air into and out of the lungs. Recurrent asthma symptoms frequently cause sleeplessness, daytime fatigue, reduced activity levels and school and work absenteeism. Asthma has a relatively low fatality rate compared to other chronic diseases. The World Health Organization estimates that 300 million people currently suffer from asthma (WHO, 2021).

2. *Chronic Obstructive Pulmonary Disease (COPD)* is a disease characterized by progressive airflow obstruction caused by an abnormal inflammatory response (tissue destruction) in the lungs as a result of chronic inhalation of harmful particles and gases (particles reaching the lungs in gas or vapor form) (Kocabas, 2000; Er, 2009). Clinically, patients with COPD experience shortness of breath (dyspnea) and cough, productive of an excess of mucus. There may also be wheeze (Jeffery, 1998). According to the WHO data, 251 million patients have COPD and every year approximately 3.17 million persons die because of COPD in the world, reported in 2016 (WHO, 2021).

3. *Pneumonia* or medically known as pneumonia bacteria is defined as a lung infection caused by bacteria, viruses, and in rare cases, parasites. This lung infection causes inflammatory cells to accumulate in small, air-filled lung sacks known as alveoli and causes the serum to arrive in this area via blood vessels to fill these alveoli. The alveoli, which are filled with serum fluid and inflammatory cells, lose their air content and fail to fulfill their respiratory function. This means oxygen cannot reach the blood and the cells of the body effectively. If pneumonia is widespread, respiratory failure may occur in the patient (TORAKS Derneği Eğitim Kitapları Serisi, 2004; Er, 2009). According to WHO data, every year approximate 2.56 million persons die because of pneumonia, reported in 2017 (WHO, 2021).

4. *Tuberculosis* is caused by a microbe known to exist for thousands of years: *Mycobacterium tuberculosis*. After the tuberculosis microbe enters the body, it can remain dormant for a long time without triggering the disease. During this period, the defense responses formed by the body keep the microbes in an inactive state. These microbes generally congregate in the lungs and the disease is considered to be very serious (Er, 2009). TB is a major cause of illness

and death worldwide and globally, 10 million new cases and 1.4 million deaths from tuberculosis occurred in 2019 (WHO, 2021).

5. *Lung cancer* is a disease in which cells in the lung tissues proliferate uncontrollably. This uncontrolled proliferation can cause these cells to invade the surrounding tissues or to spread to organs other than the lung (metastasis). The vast majority of primary lung cancers are carcinomas of the lung, derived from epithelial cells. The global cancer burden is estimated to have risen to 18.1 million new cases and 9.6 million deaths in 2018. One in five men and one in six women worldwide develop cancer during their lifetime, and one in eight men and one in 11 women die from the disease. Worldwide, the total number of people who are alive within five years of a cancer diagnosis, called the five-year prevalence, is estimated to be 43.8 million (Er, 2009; Er et al., 2010; WHO, 2021).

5.3 CONVOLUTIONAL NEURAL NETWORK (CNN)

The convolutional neural network is a type of feed-forward artificial neural network whose connection between neurons is inspired by the visual cortex of animals. CNN is among the most popular deep learning algorithms and it learns to perform the classification process directly from image, video, text, or audio files. CNN is quite similar to ordinary ANNs, and it consists of neurons with learnable weight and bias values just like ordinary ANNs (Er et al., 2012; Lu et al., 2017; Aghdam and Heravi, 2017). The biggest difference of CNN from ordinary ANNs is that by nature it assumes its inputs as two- or three-dimensional images. This situation causes a significant reduction in the number of network parameters, and at the same time, it prevents overfitting in problems with a high number of features and a low amount of data, thereby increasing efficiency. The main reason for using the CNN architecture in this study is that it automatically extracts features while learning. With this aspect, it provides more advantageous and successful results than ordinary machine learning algorithms.

5.3.1 ARCHITECTURE OF A TYPICAL NEURAL NETWORK VERSUS CNN

To understand CNN architecture, it is necessary to understand the operation of artificial neural network architectures. Ordinary ANNs compute the result by processing an input applied as a single vector through a series of layers. While each layer consists of a series of neurons, every neuron in each layer is fully connected to all other neurons in the layers before and after it with certain weight values (fully-connected). Considering that we have a data set consisting of 32 × 32 grayscale images, the input vector of the ANN performing any classification process designed for this data set will be 1024 × 1(32 × 32). Since all neurons in a fully connected layer will be connected to all neurons in the next layer, for our sample, we will have 1,024 parameters for each neuron in the first hidden layer following the input layer. Due to the nature of the fully connected architecture, any increase in the depth of the ANN (the number of layers) or the number of neurons in each layer will significantly increase the number of parameters, which will, in turn, cause an over-fitting of the problem. In other words, the ANN that has been adapted too much to the training set will not be able to correctly classify the samples, except for those in the training set.

CNN can be trained quickly with its special architecture that is well adapted to image classification problems. Unlike ordinary ANNs, CNNs neurons are arranged in three dimensions. As in ordinary ANNs, each neuron in the input layer of CNNs is connected to the neurons in the hidden layer, but unlike the completely connected structure in ordinary ANNs, each input neuron is not connected to all neurons in the hidden layer. Instead, each neuron in the hidden layer is connected to a small region (for example, a 3 × 3 area) in the input layer. This region in the input layer (input image) is called the local receptive area. There is a neuron in the hidden layer for each local receptive area in the input image. The response of each local receptive area is calculated using the convolution process. Since these weights are also used by local receptive areas while connecting to the neurons in the hidden layer, they are also called shared weights.

In ordinary ANNs, each value belonging to the input is transmitted to the neuron in the next layer after being multiplied with different weight values. However, in CNNs, a convulsion filter progresses at a certain pace over the input to create a map of the features. Thanks to the shared weight architecture, it decreases the number of CNN parameters considerably compared to ordinary ANNs, and this enables us to obtain very effective solutions.

As it can also be seen in Figure 5.2, the extraction of features is carried out by experts to select suitable inputs for the data set. Although there are different methods that determine the effectiveness or redundancy of these features (Principal component analysis and independent component analysis), deep learning methods determine this feature extraction process without the need for an expert in architecture (Olmez, 2020). The input layer holds all raw values of the applied input. The dimensions of the input layer are determined by the data set to be used in the CNN.

As seen in Figure 5.3, a typical convolutional neural network has three types of layers that are repeated in different numbers and combinations between the input layer and the fully connected layer. These layers are the convolution layer, the ReLU layer, and the Pooling layer, respectively. While creating a CNN, these three types of layers are repeated over and over to adjust the depth of the network.

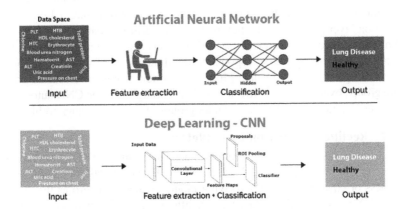

FIGURE 5.2 Comparison of CNN and ANN architectures.

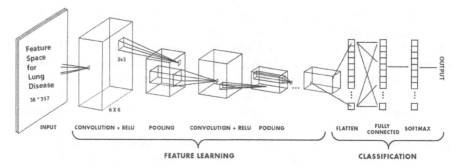

FIGURE 5.3   Feature extraction and classification layers in the CNN architectures (Olmez et al., 2020).

5.3.1.1   Convolutional Layer

The purpose of the convolution layer in a CNN is to extract features by processing the input with the convolution filters that are integrated within it. The inputs and outputs of the convolution layer are in the form of geometric objects (tensors) in which the multidimensional data are symbolized. The process of convulsion is a type of summation of products and it is used to calculate the outputs of the receptive areas in the input layer. The dimensions of the convulsion filter to be specified determines the dimensions of the receptive area and this filter is passed over the input of the convulsion layer to calculate the outputs of each receptive area.

There are three hyper-parameters to be selected for the convulsion layer during the design of the network. These are the dimensions of the convolution filter, the size of the strides while the convolution filter is being passed over the input and whether any padding will be applied to the input or not (Convolution Operations, 2020).

The convolution layer is the basic building block of CNN. The parameters of the layer are the values of the filters of particular dimensions that represent the receptive fields on the input. Convolution filter values (parameters) are adjusted during network training, enabling the network to successfully perform the relevant classification process. By increasing the number of convolution filters, the number of features to be removed from the input can be increased. As shown in Figure 5.4, filters in the convolution layer pass over the input to obtain the resulting matrix.

While the convolution layer of the CNN extracts a single feature for each receptive area on the image that corresponds to the dimensions of the convolution filter when working on images, in data processing, a new feature is extracted for each feature group covered by the convolution filter. Through this, the CNN automatically defines new features from different feature groups in data processing problems.

5.3.1.2   Rectified Linear Unit Layer (ReLU)

The ReLU layer is the layer that applies an activation function of $f(x) = \max(0, X)$ to each element in its input. ReLU, which is a nonlinear activation function, sets any inputs less or equal to zero while leaving inputs greater than zero as they are. In a typical CNN model, the ReLU layer is used after the convolution layers. As it can be seen in Figure 5.5, the ReLU layer is applied separately for each element of its input

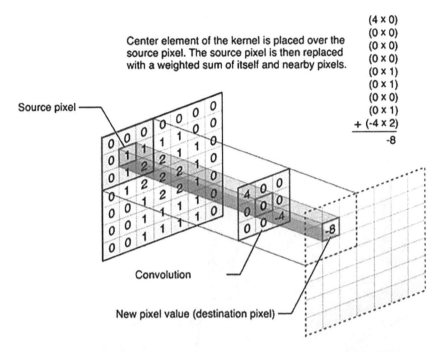

FIGURE 5.4 Convolution process.

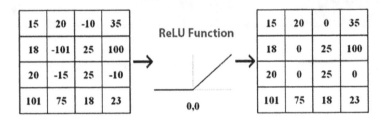

FIGURE 5.5 ReLU process.

and it has equalized values smaller than 0 to 0 while leaving any values greater than 0 as they are.

The use of the ReLU function is preferred because it does not create a significant correlation in the generalization accuracy in comparison to other activation functions (such as sigmoid or hyperbolic tangent), despite being several times faster than they are. This difference provides great ease of application in deep neural networks where the computational load is quite high.

5.3.1.3 Pooling Layer

Another important concept for CNNs is the pooling layer which performs non-linear down sampling. The pooling layer is a non-parametric layer, and it does not contain any parameters. Although several different functions are used for pooling, the most common one is the max-pooling function. This process divides its input into

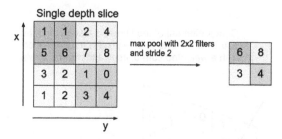

FIGURE 5.6 Max pooling process with 2 × 2 filter (CS231n: Convolutional Neural Networks for Visual Recognition, 2020).

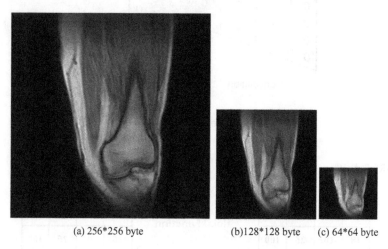

(a) 256*256 byte   (b)128*128 byte   (c) 64*64 byte

FIGURE 5.7 Sample max-pooling process on MR images (Olmez et al. 2020).

non-overlapping rectangles and for each sub-region, it produces the highest value of the region in question as output. The rule that is in effect here is that once a feature has been determined, its exact location is not as important as its rough location in comparison to other features. The function of the pooling layer is to decrease the spatial dimensions of the input gradually in order to reduce the number of parameters and computations, thereby controlling overfitting. Figure 5.6 shows a max-pooling process performed with strides of 2 units using a 2 × 2 filter. The 16-pixel field of view was reduced to four pixels after the max-pooling process.

In Figure 5.7, a series of MRI images in 8-bit depth and 256 × 256 resolution were subjected to max-pooling process twice with a filter of 2 × 2 using strides of two. As a result, a 16-fold reduction was achieved.

5.3.1.4 Fully Connected Layer

The fully connected layer is the last one in a CNN architecture, and it generates class scores. Following the convolution and pooling layers, the final high-level extraction is performed in the fully connected layer. After applying a series of convolution and pooling processes to a typical CNN input, an output in the form of a three-dimensional tensor is obtained. These outputs are transformed into a flat vector and transferred to the fully connected layer. The input size of the fully connected layer is determined by the previous layer and its output is determined by the number of classes in the data set. Each neuron on the fully connected layer is connected to each output neuron with specific weight values. This type of connection is the same as the fully connected architecture of ordinary ANNs (Kingma et al., 2015).

5.3.1.5   Softmax Function

Sometimes appearing as a separate layer in the literature, the Softmax function is used very often in the outputs of deep learning models. It has a similar structure to the Sigmoid function and good classification performance. The Softmax function sets the class scores generated in the fully connected layer to probability-based values between 0 and 1. The Softmax function takes an N-dimensional input vector and produces a secondary N-dimensional vector, each element of which is given a value between 0 and 1. Although the Softmax function is generally used in the output layer of deep learning models, other classifiers, such as the Support Vector Machine (SVM), can also be used. Since it is an exponential function, the Softmax function makes the difference between classes even more explicit.

5.3.2   TRAINING OF CNN

A CNN with appropriate parameter values (weights and bias) will correctly assign a given input to one of the classes it has in its output. The purpose of CNN training is to find the most appropriate CNN parameters to perform the correct classification process for a classification problem. At the start of the CNN training, the values of weights and biases are assigned random values or the process may be started using the parameter values of a different neural network performing similar tasks, which is a method that we call transfer learning. In reality, we can compare transfer learning to some kind of experience transfer. If parameters are received with a transfer learning from a suitable CNN performing a similar task, CNN training can be performed with less data and a smaller number of iterations. High amounts of data and computation required for the training of deep neural networks can be alleviated with transfer learning (Olmez, 2020).

After the initial values of parameter values are determined, each sample in the training set is fed as input to the CNN to calculate class scores generated for each class in its output. By applying these generated class scores to a cost function, whether the CNN produces appropriate results with the existing parameters or not is determined. If our parameter values are not suitable in terms of intuitiveness, our cost function will generate relatively high values. In essence, training a CNN is the process of finding CNN parameters that minimize the cost function (Olmez, 2020).

In the literature, the concepts of cost function and loss function are often confused and used interchangeably. While the loss function calculates the error for a single sample in the training set, the cost function calculates the total error for the entire training set. The cost function is sometimes termed as the loss layer in the literature.

By using the training set during CNN training, the cost function calculates how much the class scores produced by the CNN with the current parameters deviate from the actual values they need to be. The cost function is taken into account in determining how the CNN parameters will be updated in the next training step.

For a single class NN classifier with a single neuron at its output layer, the cost function calculates the cost of each sample in the training set at the NN output. For a multi-class NN classifier with multiple neurons at its output layer, the cost function calculates the cost for all classes at the NN output for each sample in the training set. Cost values for each sample are generated based on the output class they belong to. Moreover, cost values for classes that they do not belong to are also generated. For intuitive multi-class classification, the output value for the class of an input sample should be as large as possible. For the example provided, the cost value is expected to decrease as the output value of the class increases. NN is expected to generate smaller values at the output for classes that the input value does not belong to, and it is anticipated that for these classes, cost values should decrease as the output decreases. In this way, the costs generated by each sample for all classes are summed up and divided by the number of samples, thereby calculating the average cost (Olmez, 2020).

NN training is the finding of network parameters that minimize the cost function for the training set. The backpropagation algorithm used in the training of NNs is an algorithm that minimizes the cost function. The partial derivatives obtained in the backpropagation algorithm are used together with the gradient descent or a more advanced optimization algorithm to find network parameters that minimize the cost function. For the samples in the ANNs training set, the backpropagation algorithm compares the values produced by the ANN in the output layer with the intended values to determine its output error. Afterwards, this output error is propagated from the last layer to the first layer by taking partial derivatives.

After establishing a cost function that calculates the total error with the training set for CNN training, the next step is to establish a training algorithm that finds CNN parameters that will minimize this cost function. The gradient descent algorithm, which is frequently used in training CNNs, performs this optimization by advancing towards the negative of the gradient of the cost function with a certain learning rate.

The gradient descent algorithm is a cyclical process, and it is subtracted from the parameters by multiplying the cost function obtained using the current parameters gradient by a certain learning rate. Thus, new parameter values are obtained. The algorithm is continued with these new parameter values. As shown in Figure 5.8, as the derivatives of the cost function become smaller with each new version of the parameters, the learning steps also become smaller. Since the derivative of the cost function will be equal to 0 when the cost function demonstrated in Figure 5.8 reaches its minimum point, the updated value of the gradient descent algorithm will be equal to its current value. At this point, the algorithm will be terminated. If a very large value is selected for the learning rate α, the gradient descent algorithm may end up constantly missing the minimum point since the learning step will also become too large. If a very small value is selected for the learning rate, the time that the value takes to reach a minimum will be increased considerably since the learning steps will be rendered too small. Therefore, an optimum value should be selected for the learning rate α (Olmez, 2020).

Training sets used in an ordinary deep learning application may comprise thousands of data. In such applications, updating the parameters by calculating the total

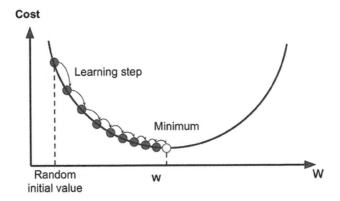

FIGURE 5.8 Gradient descent algorithm.

error over the entire training set can be a heavy computational burden. While the process of updating the parameters over the training set in each iteration is called the Batch gradient descent, the process of dividing the training set into smaller batches (mini-batches) and updating the parameters with gradient descent algorithm over these batches is called mini-batch gradient descent (Lu et al., 2017). The process of updating the parameters over a single batch with the algorithm is called an iteration, and the process of passing all batches (i.e., the entire training set) over the algorithm a single time is called an epoch (MATLAB, 2020).

The standard Gradient descent algorithm uses a single learning rate to update the weights. In the literature, there are optimization algorithms such as AdaGrad, RmsProp, AdaDelta, and Adam, which can be considered as extensions of the gradient descent. These algorithms can automatically adjust the learning rates by themselves. Adam is the method recommended by Kingma et al., (2015) and they have demonstrated that it produces better results than other optimization algorithms.

5.4 LUNG DISEASE DIAGNOSIS USING CNN

5.4.1 Data Description and Feature Extraction

In this study, an automatic diagnosis system with a CNN has been proposed for the detection of various lung diseases. The data set used in the study consists of 38 different features of 357 subjects. These features are (Laboratory examination): complaint of cough, body temperature, ache on chest, weakness, dyspnoea on exertion, rattle in chest, pressure on chest, sputum, sound on respiratory tract, habit of cigarette, leucocyte (WBC), erythrocyte (RBC), trombosit (PLT), hematocrit (HCT), hemoglobin (HGB), albumin2, alkalen phosphatase 2 L, alanin aminotransferase (ALT), amylase, aspartat aminotransferase (AST), bilirubin (total + direct), CK/creatine kinase total, CK–MB, iron (SERUM), gamma–glutamil transferase (GGT), glukoz, HDL cholesterol, calcium (CA), blood urea nitrogen (BUN), chlorine (CL), cholesterol, creatinin, lactic dehydrogenase (LDH), potassium (K), sodium (NA), total protein, triglesid, uric acid (Er, 2009).

The 38 features were included in the epicrisis reports of patients who are hospitalized due to lung disease. This study was carried out with the original data that were obtained and determined by the specialist physician using epicrisis reports. In the feature determination phase, all possible measurements were taken into consideration (physical, blood and urine analysis), and the selection of effective features is left to the CNN architecture.

Out of the entirety of the subjects in the data set, 100 had no lung diseases whatsoever, while 257 were afflicted by six different lung diseases. Due to its architecture, the CNNs accept inputs as 2- or 3-dimensional images. In this study, 36 of the 38 features of the subjects were converted into a 2-dimensional matrix with 6 × 6 dimensions, which was then applied to a CNN. First, the data combination with the highest success rate was determined for the training set by individually training the designed CNN with the inputs obtained with all combinations of 36 features out of 38 features in the CNN data set. Afterwards, changes were made in the optimization algorithm, hyper-parameters, and network architecture to determine the most suitable CNN model for the data set.

5.4.2 Applied CNN Architecture

As shown in Figure 5.9, the CNN architecture originally designed for lung disease classification includes input in the form of a 6×6 matrix and two outputs. Following the input layer, one convolution layer, one max-pooling layer, and one ReLU layer were used, respectively.

In order for the classification of lung disease to be made successfully with current data, the hyper-parameters should be selected appropriately. This process was applied by using filters of different sizes (2 × 2, 3 × 3, … N × N) in the convolution layer and max pooling layer. All possible filter sizes were tested for this study. Thanks to the testing process, the most suitable filter was determined for convolution and max-pooling layers. At the end of this process, the most appropriate feature extraction was achieved. In the convolution layer, 30 3 × 3 convolution filters were used and, in the max-pooling layer, 2 × 2 max pooling filters were used with two strides. In this architecture, after applying a 2 × 2 max pooling filter with two strides to the feature map obtained in the classification layer of the CNN with 30 3 × 3 filters, 120 features were obtained in the fully connected (FC) layer.

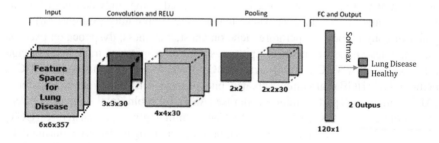

FIGURE 5.9 CNN architecture and parameters designed for lung diseases.

In designing the architecture, it was attempted to find the most successful set of combinations in the current feature set for the detection of lung diseases by quickly designing a CNN. The dataset used in CNN training was divided into two, with 70% set aside for training and 30% for testing. Likewise, all combinations of 36 of 38 input features were used to perform 703 different training with the divided training set. The purpose of this process is to prevent the selection of a low success feature combination from the feature set. The training was performed using Batch Gradient Descent with a learning rate of 0.01, and 3,000 was selected as the maximum number of epochs. Upon completion of the selection of the feature set, some alterations were made to the CNN architecture during the next step. Afterwards, it was attempted to determine the most successful model.

Following the determination of the feature set to be used for CNN training, network training was carried out with different hyper-parameters, and the most suitable optimization algorithm and hyper-parameters were selected. At this stage, the classical gradient descent algorithm was compared with the Adam algorithm, and it was tried in various combinations with different learning rates and batch normalizations. The batch-normalization method suggested by Ioffe and Szegedy (Ioffe et al., 2015) was determined to increase the speed of the CNN training and also the rate of success. Batch normalization was applied between the convolution layer and the non-linear activation function in the CNN and it was performed by normalizing the output values of the convolution layer.

5.4.3 RESULTS

In this section, the training process was performed 10 times by dividing the dataset during each training randomly into 30% for test and 70% for training. The results obtained are provided in Table 5.1 below. On average, the most successful result was

TABLE 5.1
Results Obtained by Dividing the Data Set Randomly by 30% Testing and 70% Training

Experiment	A	B	C	D	E	F	G	H
Optimizer	GD	Adam	GD	Adam	GD	GD	Adam	Adam
LR	0,001	0,001	0,001	0,001	0,0001	0,0001	0,0001	0,0001
Epoch	3000	3000	10,000	10,000	10,000	10,000	10,000	10,000
BatchNorm	-	-	var	var	-	var	-	var
1	89,72	86,92	90,65	88,79	91,59	89,72	88,79	90,65
2	87,85	92,52	94,39	94,39	93,46	87,85	93,46	95,33
3	91,59	89,72	90,65	90,65	90,65	83,18	91,59	90,65
4	88,79	87,85	87,85	90,65	85,98	85,98	87,85	87,85
5	88,79	92,52	92,52	90,65	92,52	87,85	91,59	96,26
6	88,79	91,59	87,85	89,72	88,79	88,79	88,79	89,72
7	92,52	93,46	93,46	91,59	92,52	87,85	92,52	93,46
8	85,05	86,92	93,46	88,79	91,59	88,79	88,79	90,65
9	90,65	89,72	90,65	89,72	91,59	86,92	91,59	90,65
10	90,65	89,72	91,59	92,52	93,46	87,85	88,79	92,52
Mean	89,44	90,09	91,31	90,75	91,22	87,48	90,38	91,77

91.77%, which was obtained with a learning rate of 0.0001, 10,000 epochs, and using the Adam algorithm and batch normalization.

After deciding on the optimization algorithms and hyper-parameters to be used in the training, it was attempted to increase the rate of success by making various changes to the network architecture. The experiments made here were carried out on the existing architecture by reducing the size and number of the convolution filters, and with and without the pooling layer. In 10 different training experiments carried out in this manner, the testing set was randomized before each training. As a result of 10 different trainings, the highest success value obtained was 95.51%, which was achieved with architecture with 20 2 × 2 convolution filters and without applying the max-pooling process. All the results obtained are shown in Table 5.2 and the model with the highest success rate is shown in Figure 5.10.

TABLE 5.2
Results of the Study by Reducing the Size and Number of the Convolution Filters

Experiment	0	B	C	D	E	F
max-pool	2 × 2	–	2 × 2	–	2 × 2	–
conv layer	3 × 3 × 30	3 × 3 × 30	2 × 2 × 30	2 × 2 × 30	2 × 2 × 20	2 × 2 × 20
1	92,52	94,39	95,33	94,39	92,52	95,33
2	91,59	92,52	88,79	92,52	91,59	95,33
3	89,72	91,59	85,05	91,59	81,31	92,52
4	94,39	92,52	92,52	92,52	90,65	92,52
5	95,33	99,07	94,39	98,13	89,72	98,13
6	93,46	94,39	88,79	97,20	85,98	98,13
7	93,46	92,52	91,59	92,52	88,79	95,33
8	93,46	96,26	91,59	97,20	86,92	96,26
9	92,52	93,46	88,79	91,59	91,59	93,46
10	93,46	94,39	88,79	96,26	85,98	98,13
Mean	92,99	94,11	90,56	94,39	88,51	**95,51**

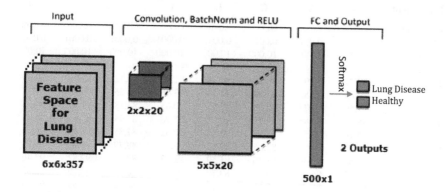

FIGURE 5.10   The model with the highest success rate.

TABLE 5.3
The Classification Accuracies Obtained by CNN and PNN Structures for Lung Diseases

CNN Structure (%)	PNN Structure (%)
95.51	91.25

In order to analyze the performance of the CNN architecture, the dataset was tested with classical machine learning methods using same database. For this purpose, a probabilistic neural network (PNN) was used for the lung diseases diagnosis. The PNN structure (Speckt, 1990; Cetin and Temurtas, 2020) used in this study has a multilayer structure consisting of an input layer, a single hidden layer (radial basis layer), and an output layer (competitive layer). In this system, real valued input vector is the features' vector, and two outputs are index of two classes (lung disease and normal). All hidden units simultaneously receive the 38-dimensional real valued input vector. The classification accuracies obtained by CNN and PNN structures for lung diseases were presented in Table 5.3.

From the Table 5.3, it can be seen that the results obtained using CNN structure were better than the results obtained using PNN structure. So, it can be said that the CNN structure can be used for the detection of lung disease. As a result, it was found appropriate to create a CNN-based decision support system on this structure.

5.5 CONCLUSION AND EVALUATION

In this study, the CNNs, which have become almost a standard in image classification, were used for data classification. In this classification, which was carried out for the detection of lung diseases, CNN models that allow good results in prediction performance by effectively using multi-dimensional data were benefitted from. In the studied CNN models, in which different variations were tested in modeling the convolutional processes in a problem-specific manner, the highest rate of success was achieved in the model with the model where the CNN filter size was the smallest and pooling was not applied. The best CNN structure was also compared with a probabilistic neural network. The results indicated that CNN structure design for the lung disease classification problem outperformed the PNN approach using predictive accuracy rate as a performance indicator.

The primary cause behind this situation is the fact that CNN can extract a higher number of features from the input matrix from a data set with a small number of features using a convolution filter with smaller filter dimensions. Besides, the application of a down-sampling pooling layer decreases the number of available features, thus reducing the success rate for this data set, which contains relatively few features.

Using convolution layers and pooling layers consecutively for data sets with a higher number of features will help to produce more successful models, as this will reduce the number of parameters. Achieving an average success rate of 95.51% with

36 features and 357 samples, a size which can be considered very small for a CNN, is very encouraging for the design of models with higher number of features and data samples.

REFERENCES

Aghdam H.A. & Heravi E.J., *Guide to Convolutional Neural Networks: A Practical Application to Traffic-Sign Detection and Classification*, Springer, 1st edn., 2017.

Arena P., Basile A., Bucolo M. & Fortuna L. Image processing for medical diagnosis using CNN. *Nuclear Instruments and Methods in Physics Research Section A: Accelerators, Spectrometers, Detectors and Associated Equipment*, 497(1), 174–178, 2003.

Ashizawa K., MaCMahon H., Ishida T., Nakamura K., Vyborny C.J., Katsuragawa S. & Doi K. Effect of an artificial neural network on radiologists' performance in the differential diagnosis of interstitial lung disease using chest radiographs. *AJR. American Journal of Roentgenology*, 172(5), 1311–1315, 1999.

Cetin O. & Temurtas F. A comparative study on classification of magnetoencephalography signals using probabilistic neural network and multilayer neural network. *Soft Computing*, 2020.

Convolution Operations, https://developer.apple.com/library (accessed November 01, 2020).

CS231n: Convolutional Neural Networks for Visual Recognition, *Stanford university course notes*, https://cs231n.github.io/convolutional-networks/ (accessed November 01, 2020).

Er O. Thoracic Disease Diagnosis using Flexible Computing and Bioinformatics Computing System. PhD Thesis, Sakarya University, 2009.

Er O., Yumusak N. & Temurtas F. Chest disease diagnosis using artificial neural networks, *Expert Systems with Applications*, 37(12), 7648–7655, 2010.

Er O., Tanrikulu A.C., Abakay A. & Temurtas F. An approach based on probabilistic neural network for diagnosis of Mesothelioma's disease. *Computers & Electrical Engineering*, 38(1), 75–81, 2012.

Gao F., Wu T., Li J., Zheng B., Ruan L., Shang D. & Patel B. SD-CNN: A shallow-deep CNN for improved breast cancer diagnosis. *Computerized Medical Imaging and Graphics*, 70, 53–62, 2018.

Geng Y., Ren Y., Hou R., Han S., Rubin G.D. & Lo J.Y.. 2.5 D CNN model for detecting lung disease using weak supervision. In Mori K, Hahn H K (eds). *Medical Imaging 2019: Computer-Aided Diagnosis*. International Society for Optics and Photonics, San Diego, California, United States, (Vol. 10950, pp. 1095030). (2019, March).

Hattikatti P., *"Texture based interstitial lung disease detection using convolutional neural network," 2017 International Conference on Big Data, IoT and Data Science (BID)* Pune pp. 18–22, 2017.

Ioffe S. & Szegedy C. *Proceedings of the 32nd International Conference on Machine Learning*, Lille, France: PMLR 37 pp. 448–456, 2015.

Jeffery P.K. Structural and inflammatory changes in COPD: A comparison with asthma. *Thorax*, 53(2), 129–136, 1998.

Khobragade S., Tiwari A., Patil C.Y. & Narke V. *Automatic detection of major lung diseases using Chest Radiographs and classification by feed-forward artificial neural network.* In *2016 IEEE 1st International Conference on Power Electronics, Intelligent Control and Energy Systems (ICPEICES)* Delhi, India pp. 1–5, 2016.

Kido S., Hirano Y. & Hashimoto N., *"Detection and classification of lung abnormalities by use of convolutional neural network (CNN) and regions with CNN features (R-CNN),"* In *2018 International Workshop on Advanced Image Technology (IWAIT)* Chiang Mai pp. 1–4, 2018.

Kingma D.P., Ba J. & Adam A. *Method for stochastic optimization*. In *Proceedings of 3th International Conference on Learning Representations*San Diego, CA pp. 1–15, 2015.

Kocabas A. KOAH: Epidemiyoloji ve Doğal Gelişim. In Umut S, Erdinç E (eds). *Kronik Obsrüktif Akciğer Hastalığı*. Ankara: Toraks Derneği Yayınları, pp. 8–25, 2000.

Liang Z., Powell A., Ersoy I., Poostchi M., Silamut K., Palaniappan K, ... & Huang J.X. *CNN-based image analysis for malaria diagnosis*. In *2016 IEEE International Conference on Bioinformatics and Biomedicine (BIBM)* Shenzhen, China pp. 493–496, 2016.

Lu L.E., Zheng Y., Carneiro G. & Yang L. *Deep learning and convolutional neural networks for medical image computing: Advances in Computer Vision and Pattern Recognition*, Springer, 2017.

MATLAB for Artificial Intelligence, https://www.mathworks.com/solutions/deep-learning/convolutional-neural-network.html (accessed November 01, 2020).

Oh S.L., Ng E.Y., San Tan R. & Acharya U.R. Automated diagnosis of arrhythmia using combination of CNN and LSTM techniques with variable length heart beats. *Computers in biology and medicine*, 102, 278–287, 2018.

Olmez E. Automatic segmentation of meniscus in MRI using deep learning and morphological image processing, Mechatronics Engineering, PhD Thesis, Yozgat Bozok University, 2020.

Olmez E., Akdogan V., Korkmaz M. & Er O. Automatic segmentation of meniscus in multi-spectral mri using regions with convolutional neural network (R-CNN). *The Journal of Digital Imaging*, 33, 916–929, 2020.

Sen I, Hossain M.I., Shakib M.F.H., Imran M.A. & Al Faisal F. In depth analysis of lung disease prediction using machine learning algorithms. In Bhattacharjee A, Borgohain S Kr, Soni B, Verma G, Gao X-Z (eds) *International Conference on Machine Learning, Image Processing, Network Security and Data Sciences*. Singapore: Springer, pp. 204–213. 2020.

Speckt D.F. Probabilistic neural networks. *Neural Networks*, 3, 109–118, 1990.

TORAKS Derneği Eğitim Kitapları Serisi – Zatürree: Nedir? Nasıl Korunulur?, Sayı A4, Turgut Yayıncılık, 2004.

Yılmaz T.F.. Astımlı Hastalarda Eğitimin Semptom Kontrolüne, Atak Sıklığına ve Yaşam Kalitesine Etkisi. İstanbul: Marmara Üniversitesi Sağlık Bilimleri Enstitüsü, 2012.

Wimmer G., Vécsei A. & Uhl A. *CNN transfer learning for the automated diagnosis of celiac disease*. In *2016 Sixth International Conference on Image Processing Theory, Tools and Applications (IPTA)* pp. 1–6, 2016.

World Health Organization, https://www.who.int/ (accessed January 01, 2021).

6 Deep Learning Methods for Diagnosis of COVID-19 Using Radiology Images and Genome Sequences
Challenges and Limitations

Hilal Arslan
Yıldırım Beyazıt University, Turkey

Hasan Arslan
Erciyes University, Turkey

CONTENTS

6.1 INTRODUCTION

Severe Acute Respiratory Syndrome Corona Virus 2 (SARS-CoV-2) has caused the coronavirus disease 2019 (COVID-19), which was first detected in Wuhan, China in December 2019 and then spread to many countries around the world (Arabi et al. 2020). The first mortality of COVID-19 was declared in January 2020, and the World Health Organization (WHO) declared that COVID-19 disease was a pandemic

(Wolfelet al. 2020). The pandemic has had significant outcomes, including many deaths. As of November 1, 2020, the WHO reported that there were nearly 46 million confirmed cases and 1.2 million deaths, globally. Moreover, the report published by WHO evaluated deaths with respect to age. Deaths were reported in 0–14, 25–64, and 65 years and over and rates of the deaths were 0.2%, 25%, and 75%, respectively. It was also reported that there was a remarkable difference between sexes and 59% of the deaths were males. According to the report by WHO, the most common symptoms of the disease are cough, fever, gastrointestinal and musculoskeletal symptoms as well as loss of taste or smell. Shortness of breath and chest pressure or pain are among the less common symptoms. Because these symptoms are like the common flu, it is difficult to conduct an early diagnosis of the virus. Therefore, it is essential to identify positive cases quickly since it spreads rapidly and poses a threat to the public health system. Furthermore, once a patient is diagnosed as being infected, some precautionary measures should be taken to slow down the spread of COVID-19.

The reverse transmission polymerase chain reaction (RT-PCR) is a standard technique used for testing the novel coronavirus (Anika et al. 2020). In some studies, the sensitivity and accuracy of the RT-PCR test used for the diagnosis of COVID-19 has been criticized (Holshue et al. 2020, Jiang et al. 2020). Although RT-PCR testing is a commonly used method for detecting COVID-19 disease, it produces a great number of false negatives. Although there are several reasons for false negatives, two main reasons are the disease stage and sampling methods causing the delay of an early diagnosis. Moreover, Holshue et al. (2020) detected SARS-CoV-2 RNA in the stool sample of a person with symptoms although the serum sample was tested as negative, repeatedly. For this reason, RT-PCR tests are not sufficient alone to detect COVID-19 disease status (Wang et al. 2020).

Another standard technique for detecting COVID-19 is Computed Tomography (CT) or chest X-ray imaging providing high sensitivity for significant clinical findings (Annarumma et al. 2019). The X-ray machine invented in 1895 by Wilhelm Röntgen provides clear monitoring of various affected areas of the body, such as bone fractures and dislocations, lung infections, pneumonia, cancers, and injuries (Rachna, 2017). In addition to providing two-dimensional images, these machines are also fast, inexpensive and easily accessible. As for the disadvantages of X-rays, the details of internal organ injuries are not clearly displayed via this technique, and radiation generated by the X-ray machine may be harmful. To overcome this limitation, the CT scan is invented by Hounsfield and Cormack in 1972. The CT scan is an advanced X-ray imaging device that produces multiple X-ray images by scanning the body 360 degrees. While the X-ray provides images of broken and dislocated bones, pneumonia as well as tumors, CT scanning provides more apparent images of soft tissues and internal organs and focuses better on the objective area than an X-ray. However, it is more expensive than an X-ray. Ai et al. (2020) compared the effect of the diagnosis value of the CT to the reverse Transcription Polymerase Chain Reaction (RT-PCR). They obtained both chest CT and PT-PCR data from 1,014 patients in Wuhan, China. Their results verified that PT-PCR has a lower sensitivity for COVID-19 prediction, and thus the chest CT may be regarded as a simple tool for COVID-19 prediction.

Medical practitioners need computer-aided techniques to detect COVID-19 cases from X-ray images and deep learning techniques to analyze the images in a state-of-the-art manner, and present outstanding performance. Furthermore, these techniques have a crucial role in the early diagnosis of COVID-19 to prevent the spread of coronavirus (Suzuki 2017, Anwar et al. 2018, Liu et al. 2018, Shi et al. 2020, Ting et al. 2020).

In this chapter, we review deep learning methods for diagnosing COVID-19 cases under two groups, which are studies using gene sequences data and studies using radiology image data. The remaining parts of this chapter are organized as follows. In Section 6.2, we provide an overview of deep learning methods used for COVID-19 prediction. In Section 6.3 and 6.4, we review studies for predicting COVID-19 using genome sequences and radiology images, respectively. In Section 6.5, we discuss challenges and limitations. Finally, Section 6.6 presents the conclusions.

6.2 DEEP LEARNING MODELS

Advances in computational power and the use of GPUs have made great progress in the development of artificial neural network (ANN) applications. These ANN applications and more complicated models such as the recurrent neural networks and convolutional neural networks are introduced. Deep learning methods present significant performance improvements in many applications. In this section, deep learning models, which are used for COVID-19 prediction, are briefly explained.

6.2.1 Convolutional Neural Networks

The convolutional neural network (CNN) (LeCun et al. 1989), which is a type of artificial neural network, is one of the well-known deep learning techniques. It is especially designed for image recognition and computer vision applications. A schematic representation of the CNN model is presented in Figure 6.1 and includes convolutional, pooling, and fully connected layers. The CNN model gets input data and applies a set of convolutional and pooling layers, and fully connected layers to them to generate an output in the end. Before pooling operations are performed, the output of each convolution operation is activated by activation functions in each layer. The convolution operation generates various numbers of feature maps according to the number of filters, which are applied. The goal of the pooling layers is to reduce the spatial dimensions of each feature map. Thus, while the informative features are

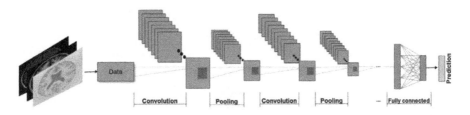

FIGURE 6.1   Schematic representation of CNN (obtained from Bernal et al. (2019)).

extracted in the feature maps, the required computational power for processing data is decreased. Finally, fully connected layers can be used for making predictions after the convolution and pooling layers act as hidden layers of ANN.

Next, we mention some of the most popular CNN architectures, which are used for designing training algorithms for image processing over the last 10 years. These are available as pre-trained models from ImageNet (ImageNet 2016). AlexNet (Krizhevsky et al. 2012) is a trained implementation of the CNN and developed by Krizhevsky, Hinton, and Sutskever in 2012. VGG Net (Simonyan and Zisserman 2014) increases depth of the CNN and consists of 16 convolutional layers as well as uniform architecture. GoogLeNet or Inception, developed by Google in 2014, has an inception architecture that is a network within a network. The beginning part of the architecture is similar to a standard convolutional network and the key part of the architecture is an intermediate layer defined as an inception module. Residual networks (ResNets) introduced by He et al. (2016) recently reached state-of-the-art performance in complex image processing and computer vision problems. They can train extremely deep networks with up to more than 1,000 layers. Further CNN architectures can be found in (Zeiler and Fergus 2014, Zheng et al. 2015, Shelhamer et al. 2017).

6.2.2 Recurrent Neural Networks

Although artificial neural networks have been commonly used in many areas, they have less memory; thus, they cannot deal with the modelling data sequencing and time dependencies in the historical models. To overcome time dependent learning, recurrent neural networks (RNN), which are a type of an artificial neural network with multiple layers, are proposed (Erman 1990). RNN includes hidden layers which are distributed over time, so they are able to save information about the past. A schematic representation of an RNN is illustrated in Figure 6.2. When an input sequence x is given, the RNN model computes the hidden sequence h, and output sequence y at time step t using Equations (6.1) and (6.2).

$$h_t = f\left(W_{x,h}x_t + W_{h,h}h_{t-1} + b_h\right) \tag{6.1}$$

$$y_t = W_{h,y}h_t + b_y \tag{6.2}$$

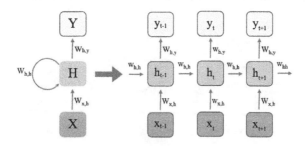

FIGURE 6.2   Schematic representation of RNN.

where $W_{x,h}, W_{h,h}$, and $W_{h,y}$ are weight matrices, b_h and b_y are bias vectors, and f is the hidden layer function. For instance, $W_{h,y}$ presents the hidden-output weightmatrix, and by presents the output bias vector.

6.2.3 LONG SHORT-TERM MEMORY

RNNs have a major drawback in that they cannot overcome the exploding or vanishing gradient problem. Moreover, they cannot deal with long-term dependencies since they include a hidden layer activation function related to the previous step only. The Long Short-Term Memory (LSTM) model (Hochreiter and Schmidhuber 1997) was designed to overcome these limitations by using memory blocks, which are a specific unit in the recurrent hidden layer. The memory blocks include self-linked memory cells and gate units holding the temporal state of the model. Every memory block includes an input gate and output gate. The input gate guards the flow of input activation into the memory cells and the output gate controls the output flow of cell activation into the rest of the network. Then, Gers et al. (2000) added a forget gate in the memory block to permit the network to reset its state. Figure 6.3 illustrates a typical LSTM block containing input gate, output gate, forget gate, the input signal x, the output y as well as the activation functions and peephole connections. Equations (6.3)–(6.7) describe the state of the gates.

$$i_t = \text{sigmoid}\left(x_t W_{x,i} + h_{t-1} W_{h,i} + c_{t-1} W_{c,i} + b_i\right) \tag{6.3}$$

$$f_t = \text{sigmoid}\left(x_t W_{x,f} + h_{t-1} W_{h,f} + c_{t-1} W_{c,f} + b_f\right) \tag{6.4}$$

$$c_t = i_t \tanh\left(x_t W_{x,c} + h_{t-1} W_{h,c} + b_c\right) + f_t c_{t-1} \tag{6.5}$$

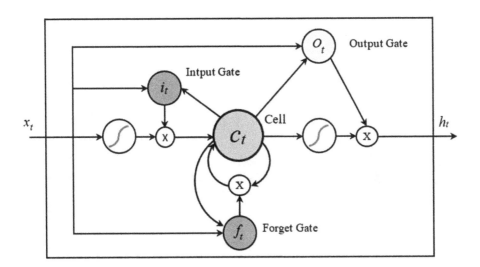

FIGURE 6.3 Schematic representation of a typical LSTM block.

$$o_t = \text{sigmoid}\left(x_t W_{x,o} + h_{t-1} W_{h,o} + c_t W_{c,o} + b_o\right) \tag{6.6}$$

$$h_t = o_t \tanh\left(c_t\right) \tag{6.7}$$

Equation (6.3) denotes the input gate i_t combining the current input x_t, the output of the LSTM unit h_{t-1} and a value of the cell c_{t-1}. In Equation (6.3), $W_{x,i}$, $W_{h,i}$, and $W_{c,i}$ refer to the weight parameters, and b_i refers to the bias parameter. Equation (6.4) denotes the forget gate, which determines which information should be removed from its previous step c_{t-1}. Thus, the activation values f_t of the forget gates at time step t are computed with respect to the current input x_t, the output h_{t-1} and the state c_{t-1} of the memory cell in the previous time step $(t-1)$. In Equation (6.4), $W_{x,f}$, $W_{h,f}$, and $W_{c,f}$ refer to the weight parameters, and b_f refers to the bias parameter. Equation (6.5) denotes cell state c_t computing based on the input gate i_t value, the input x_t, the output of the LSTM unit h_{t-1}, the forget gate f_t value, and the previous cell value c_{t-1}. In Equation (6.5), $W_{x,c}$ and $W_{h,c}$ are weights, and b_c refers to the bias parameter. Equation (6.6) denotes the output gate o_t combining the current input x_t, the output of the LSTM unit h_{t-1}, and the cell value c_t. In Equation (6.6), $W_{x,o}$, $W_{h,o}$, and $W_{c,o}$ are weights, and b_o refers to the bias parameter. Finally, the output h_t is computed by combining the current cell c_t with the current output o_t in Equation (6.7).

Recently, many studies have been proposed to enhance the performance of the LSTM model (Houdt et al. 2020). Su and Kuo (2019) proposed an extended LSTM model that retains information longer than previous RNN cells. Bayer et al. (2009) trained memory cells which are represented as differentiable computational graph structures. They used a multi-objective evolutionary algorithm to optimize the computational structure of the LSTM. The fitness functions of the evolutionary algorithm represented learning capacity of networks of memory cells. Next, we mention variants of the LSTM model, which are Bidirectional LSTM, Convolutional LSTM, and the Gated Recurrent Unit (GRU).

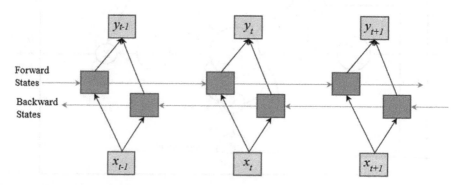

FIGURE 6.4　Schematic representation of the bidirectional LSTM.

6.2.4   BIDIRECTIONAL **LSTM**

A bidirectional LSTM is an improved version of the LSTM method. In the classical LSTM model, the current state is computed by using only one direction and does not consider information, which is possessed in the future. Bidirectional LSTM (Schuster and Paliwal 1997) was proposed to overcome this limitation and obtain more accurate state reconstruction by using information about the past and future. In this model, hidden neurons are separated into forward and backward states where neurons in the backward states are not connected to the neurons in the forward states and vice-versa. The forward, backward, and the output layers are computed using Equations (6.8)–(6.10), respectively. Schematic representation of the model is shown in Figure 6.4.

$$\overrightarrow{h_t} = \text{sigmoid}\left(x_t W_{x,\bar{h}} + \overrightarrow{h_{t-1}} W_{\bar{h},\bar{h}} + b_{\bar{h}}\right) \tag{6.8}$$

$$\overleftarrow{h_t} = \text{sigmoid}\left(W_{x,\bar{h}} x_t + \overleftarrow{h_{t+1}} W_{\bar{h},\bar{h}} + b_{\bar{h}}\right) \tag{6.9}$$

$$y_t = \overrightarrow{h_t} W_{\bar{h},y} + \overleftarrow{h_t} W_{\bar{h},y} + b_y \tag{6.10}$$

6.2.5   CONVOLUTIONAL **LSTM**

The LSTM model cannot process spatio-temporal data, which is a major drawback. Convolutional LSTM (Shi et al. 2015) was proposed to overcome this problem. The input sequences, cell output sequences, hidden state sequences, input, output, and forget gates of the convolutional LSTM model are 3D tensors. The future state of a certain cell in the convolutional LSTM model can be defined with respect to the inputs and past states of its local neighbours. This is done by performing a convolution operator instead of making multiplication operator in the LSTM model (Equations (6.3)–(6.7)). The Convolutional LSTM model is represented by Equations (6.11)–(6.15).

$$i_t = \text{sigmoid}\left(x_t * W_{x,i} + h_{t-1} * W_{h,i} + c_{t-1} \circ W_{c,i} + b_i\right) \tag{6.11}$$

$$f_t = \text{sigmoid}\left(x_t * W_{x,f} + h_{t-1} * W_{h,f} + c_{t-1} \circ W_{c,f} + b_f\right) \tag{6.12}$$

$$c_t = i_t \circ \tanh\left(x_t * W_{x,c} + h_{t-1} * W_{h,c} + b_c\right) + f_t \circ c_{t-1} \tag{6.13}$$

$$o_t = \text{sigmoid}\left(x_t * W_{x,o} + h_{t-1} * W_{h,o} + c_t \circ W_{c,o} + b_o\right) \tag{6.14}$$

$$h_t = o_t \circ \tanh\left(c_t\right) \tag{6.15}$$

where "o" represents Hadamard operator and "*" represents the convolution operator.

6.2.6 GATED RECURRENT UNIT

A gated recurrent unit (GRU) (Cho et al. 2014) is a simple version of the LSTM model without an output gate. Thus, the GRU model includes less parameters than the LSTM model and thus is less complicated. The GRU can solve the vanishing gradient problem using only two gates which are an update gate and a reset gate. The update gate merges the input gate and forget gate and decides how much information flows into memory. On the other hand, the reset gate decides the information owing out of memory. They can be trained to hold information from the past or remove irrelevant information. The GRU model is presented in Equations (6.16)–(6.19).

$$z_t = \text{sigmoid}\left(x_t W_{x,z} + h_{t-1} W_{h,z} + b_z\right) \tag{6.16}$$

$$r_t = \text{sigmoid}\left(x_t W_{x,r} + h_{t-1} W_{h,r} + b_r\right) \tag{6.17}$$

$$\tilde{h}_t = \tanh\left(x_t W_{x,h} + \left(r_t \circ h_{t-1}\right) W_{h,h} + b_h\right) \tag{6.18}$$

$$h_t = z_t \circ h_{t-1} + \left(1 - z_t\right) \circ \tilde{h}_t \tag{6.19}$$

Equations (6.16) and (6.17) denote the update gate z_t and the reset gate r_t, respectively. They combine the current input x_t and output of the previous hidden state h_{t-1} by multiplying their own weights. In these equations, $W_{x,z}$, $W_{h,z}$, $W_{x,r}$, and $W_{h,r}$ refer to the weights, and b_z and b_r refer to the bias parameters. The current candidate hidden state $\tilde{h}_t$ is computed using Equation (6.18) and uses the reset gate to keep the relevant information from the past. The final hidden state h_t is computed using Equation (6.19), and keeps information for the current unit and puts it down in the network. A schematic representation of the GRU model is presented in Figure 6.5.

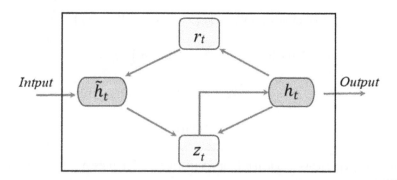

FIGURE 6.5   Schematic representation of the GRU model.

6.3   DIAGNOSIS OF COVID-19 USING GENOME SEQUENCES

The novel coronavirus genome has been entirely sequenced depending on data collected from patients, who have experienced this infectious disease around the world. Genome sequencing is extremely important for the development of optimal diagnostic methods and vaccine studies. Since the COVID-19 virus has a large mutation rate due to its nature, its genome sequencing has been highly useful and facilitated further studies in the fight against coronavirus. As the impact of this mutation on available diagnostic tests is worrying, these diagnostic tests have high probability of generating false negatives for a COVID-19 positive patient.

Methods using gene sequences data may be grouped into two parts, alignment-based and alignment-free methods. Alignment-based methods can be used to classify genome sequences especially when the size of data is small. In other words, if the size of the data becomes large, they cannot be analyzed due to long computation time. In this case, alignment free methods, machine and deep learning are preferred and can be efficiently used for many virus classification problems (Randhawa et al. 2019, Solis-Reyes et al. 2020). In this section, we review deep learning methods detecting COVID-19 using genome sequences.

Randhawa et al. (2020) combined supervised machine learning methods with digital signal processing. They applied six type supervised machine learning classifiers (linear discriminant, linear support vector machine, quadratic support vector machine, fine k-nearest neighbor (kNN), subspace discriminant, and subspace kNN) to 29 sequences of COVID-19, and 20 sequences of each of alphacoronavirus, betacoronavirus, and deltacoronavirus. They achieved 100% accuracy when the linear discriminant method was used. Naeem et al. (2020) used kNN method and the trainable cascade-forward back-propagation neural network method to classify COVID-19 cases among SARS-CoV and MERS-CoV cases. They used 76 sequences for each of COVID-19, SARS-CoV, and MERS-CoV.

Naeem et al. (2020) obtained the DNA sequences selected from a suitable site for their research needs and then converted these DNA sequence values by Gene Sequencing Program (GSP) methods into corresponding numerical values in accordance with the intended use. In the classification step, they used both deep learning and machine learning classifiers and obtained results were evaluated by using different performance measures. They evaluated results by using different performance measures, accuracy, F-measure, error rate, and Matthews correlation coefficient. Their experimental results reveal that performance of the kNN algorithm was higher than the cascade-forward neural network in all COVID-19 classification processes, and it achieved an accuracy of 100%. The cascade-forward neural network has lower performance than kNN in all the classification processes mentioned above. They numerically determined the performance of the proposed classifiers and measured accuracy, F-measure, error rate and Matthew's correlation coefficient for the kNN algorithm as 100%, 100%, 0 and 1, respectively and for the cascade-forward back-propagation neural network as 98.89%, 98.34%, 0.0111, and 0.9754, respectively. They declared that the kNN classifier gives more satisfactory results than another classifier in their study. Arslan and Arslan (2021) developed a new COVID-19 detection method by fusing kNN Classifier equipped with various distance metrics and

CpG island features extracted from complete genome sequences of human coronaviruses. They collected genome sequences of human coronaviruses from the 2019 novel coronavirus database (2019nCoVR) in the China National Bioinformation Center containing the genome sequences of different types of coronaviruses. They declared that their proposed method discriminated COVID-19 from all genome sequences of human coronaviruses an accuracy of 98.4% in a few seconds when the KNN classifier endowed with any L1 type metric is used.

6.4 DIAGNOSIS OF COVID-19 USING RADIOLOGY IMAGES

In this section, we deal with deep learning studies using medical image data (chest X-ray and CT scan). Chen et al. (2020) provided a comprehensive survey related to existing studies in the literature performed by using artificial intelligence and its variants for fighting against COVID-19 disease. They noticed that deep learning methods were successfully applied to almost every step of the struggle with COVID-19 in comparison with other coronaviruses types such as SARS-CoV in 2003 and MERS-CoV in 2012. Alakus and Turkoglu (2020) examined different COVID-19 data from Hospital Israelita Albert Einstein in Sao Paulo, Brazil using ANN, CNN, CNNLSTM, CNNRNN, LSTM and RNN deep learning models. They stated that the CNNLSTM model demonstrated better performance than all the other algorithms with an accuracy of 92.3%. Huang et al. (2020) proposed a CNN model to forecast the number of COVID-19 cases. They particularly dealt with the cities of China, where there were many COVID-19 cases. As a result of the experiments performed, they stated that the CNN model presented better results than other deep learning methods with high prediction accuracy. In a similar study introduced by Alazab et al. (2020), the authors proposed a CNN approach for automatically differentiating COVID- 19 patients from non-patients, based on a dataset including 1,000 patient chest X-ray images. Their proposed system achieved an accuracy of 90%–95%. Using three prediction methods, they also provided an intelligent model to estimate the numbers of patients who had contracted, recovered, and who would die from the disease within the next 7 days. They reported that their experimental results showed a satisfying and promising performance for the prediction of the number of those who had COVID-19 positive cases, recovered, and died with an average accuracy of 94.80% and 88.43% in Australia and Jordan, respectively. Since real-time RT-PCR test has a low positive proportion in the first stage of COVID-19 disease (Michael and Wei 2020, Xu et al. 2020) proposed an early diagnosis system by using a multiple CNN deep learning technique for discrimination of COVID-19 pneumonia cases. They extracted some distinctive characteristic features of COVID-19 that are separating from other types of viral pneumonia, such as influenza. They declared that their model reached an accuracy rate of 86.7% for classification of COVID-19, influenza, and healthy cases. Narin et al. (2020) constructed an automatic artificial intelligence system, based on CNN, for diagnosing of COVID-19 cases using three types of binary datasets including X-ray images. They implemented five pre-trained CNN- type deep neural networks which are ResNet50, ResNet101, ResNet152, InceptionV3, and Inception-ResNetV2. When the effects of the proposed CNN-based models on binary datasets under consideration were compared, they found that the most reliable and robust model in binary classification was created with ResNet-50, which had the best detection accuracy at 98%.

Being impressed by (Eman and Mohamed 2016), Ahsan et al. (2020) established a deep learning combining MLP with CNN to predict and discriminate COVID-19 cases from chest X-ray images. They here employed MLP to process the data including information of age, gender, and temperature and CNN to extract features from chest X-ray images. They used the adaptive learning rate optimization algorithm (Kingma and Ba 2017), stochastic gradient descent (Zhang et al. 2018) and root mean square propagation (Dauphin et al. 2015) algorithms to optimize and test the model. Their proposed method achieved remarkable results.

Chimmula and Zhang (2020) constructed a forecasting model to estimate the course and likely end time of the COVID-19 outbreak in Canada as well as around the world. To forecast future COVID-19 cases, they proposed LSTM networks. Based on the experimental results, it was predicted that the possible endpoint of this outbreak would be around June 2020 in Canada. Arora et al. (2020) intended to construct an effective detection model with the help of the RNN and LSTM techniques to predict the number of people suffering from COVID-19 disease. Their proposed model provided a remarkable accuracy with less than 3% error for daily forecasting and less than 8% for weekly forecasting.

Ozturk et al. (2020) applied the DarkCovidNet deep learning model to raw chest X-ray images and obtained a binary classification (COVID-19, no-findings). The authors also created triple classification (COVID-19, no-findings, pneumonia) using the DarkCovidNet classifier on the same dataset. They reported that the binary classification achieves an accuracy of 98.08% and the triple case achieves an accuracy of 87.02%. Jain et al. (2020) constructed a new intelligent prediction system to discriminate COVID-19 positive cases by analyzing chest X-ray images of healthy, bacterial pneumonia, viral pneumonia, and COVID-19 cases via the ResNet18, ResNet101, DenceNet121 and VGG-16 deep learning models. Experimental results exhibited that the ResNet101 method showed a promising and reliable result with the highest accuracy of 98.93% when compared to all other algorithms. Fu et al. (2020) constructed a classification model using ResNet-50 method for correctly classifying COVID-19 viral pneumonia from non-COVID-19 viral pneumonia, COVID-19 pneumonia, bacterial pneumonia, pulmonary tuberculosis, and normal lung cases. In order to train and validate their model, they collected a dataset consisting of 60,427 CT scan images belonging to 520 patients. Their proposed approach achieved an accuracy of 98.8%, sensitivity of 98.2%, specificity of 98.9%, true positive value 94.5% and true negative value 99.7%. Asnaoui and Chawki (2020) developed a comparative work using VGG16, VGG19, DenseNet201, Inception ResNet V2, Inception V3, Resnet50, and MobileNet V2 methods, which are novel deep learning models, for distinguishing COVID-19 cases from other viral pneumonias. They collected 6,087 chest X-rays and CT images (1,583 normal, 231 of COVID-19, 2,780 images of bacterial pneumonia, and 1,493 of coronavirus) for their experimental purpose. They used a table of confusion to measure the performances of their model. Then, they observed that both the Inception-Resnet-V2 and DenseNet201 models provided highly accurate results relative to other models in this study (they found out an accuracy of 92.18% for Inception-ResNetV2 and an accuracy of 88.09% for Densnet201).

Apostolopoulos and Bessiana (2020) employed the VGG19 and Mobile Net deep transfer learning models for detection and diagnosis of COVID-19 positive cases among 224 COVID-19, 700 bacterial pneumonia, and 504 healthy chest X-ray images. They stated that the VGG19 and the MobileNet methods had the best classification accuracy when compared to other CNN methods. They declared that the VGG19 deep learning method gave a promising and remarkable result in the classification of COVID-19 with an accuracy of 98.75%. Brunese et al. (2020) suggested a three-step approach: First, they tried to detect the presence of pneumonia in chest X-ray images. Second, they distinguished X-ray images of COVID-19 pneumonia from that of generic pneumonia. The final step was aimed at concretely determining the areas with COVID-19 symptoms in a chest X-ray image with COVID-19 pneumonia. They used a deep learning technique depending on the VGG-16 (Visual Geometry Group) method to construct their first and second models. They conducted the experiment on 6,523 chest X-ray images obtained from various institutions all over the world and then they observed that the proposed approach had an accuracy of 99% in COVID-19 detection. They stated that the average COVID-19 detection time of their proposed model was about 2.5 seconds. Hemdan et al. (2020) developed a new deep learning structure referred to as COVIDX-Net using X-ray images. The data set they used in their research contained 50 chest X-ray images consisting of 25 confirmed positive COVID-19 images and 25 non COVID-19 images. Their proposed COVIDX-Net method operated seven different deep convolutional neural network models (MobileNet, ResNet-v2, Inception-ResNetv2, Xception, Inception-v3, DenseNet, and modified VGG19). Their experimental results showed that the VGG19 and DenseNet models provided robust performances in the detection of COVID-19 negative and COVID-19 positive cases with an F-score of 89% and 91%, respectively. On the other hand, it was observed that the Inception v3 model was of the lowest performance in the classification process with F-scores of 67% for COVID-19 negative images and 0% for COVID-19 positive images.

Basu et al. (2020) developed an alternative screening method of COVID-19 called Domain Extension Transfer Learning. They generated some distinguishing features from chest X-ray images. Their proposed classification system achieved an accuracy of 95.3% in detecting COVID-19 disease. Ghoshal and Tucker (2020) augmented X-ray images of the lung from the publicly available GitHub repository provided by Dr Joseph Cohen by combining them with chest X-ray images obtained from the open source Kaggle repository. In order to estimate model uncertainty, they trained a Monte-Carlo Dropweights Bayesian deep learning classifier with the help of the transfer learning method on COVID-19 X-Ray images. Their experimental results revealed that there was a robust relationship between model uncertainty and prediction accuracy. The studies mentioned above are summarized in Table 6.1.

6.5   DISCUSSION

The availability of large-scale and high-quality data sets in deep learning algorithms plays an extremely important role in the accuracy of experimental results obtained during the establishment of the model. Deep learning studies on datasets of

TABLE 6.1

Comparison of Existing Deep Learning Studies in COVID-19 Datasets

Study	Method Used	Dataset	Image Classes	Performance Measure (Accuracy%)
Apostolopoulos and Bessiana (2020)	VGG19 700 Mobile Net 504	224 COVID-19 X-ray images, 700 images bacterial pneumonias, 504 images normal conditions	COVID-19, Bacterial pneumonia, Normal	98.75%
Fu et al. (2020)	ResNet-50	89,628 CT scan images	COVID-19 pneumonia, Non-COVID-19 viral pneumonia, Bacterial pneumonia, Pulmonary tuberculosis, Normal	98.8%
Randhawa et al. (2020)	Linear Discriminant, Linear SVM, Quadratic SVM, Fine kNN, Subspace Discriminant, Subspace kNN	29 COVID-19 virus sequences, 20 alphacoronavirus sequences, 20 betacoronavirus sequences, 20 deltacoronavirus sequences	COVID-19, other types of viruses	100%
Naeem et al. (2020)	Discrete Fourier Trans. with kNN, Discrete Cosine Trans. with kNN, Seven Moment Invariants with kNN	76 COVID-19 sequences, 76 SARS-CoV sequences, 76 MERS-CoV sequences	COVID-19, SARS-CoV, MERS-CoV	100%
Arslan and Arslan (2021)	kNN with L1 type metrics CpG based features	1,000 COVID-19 sequences, 592 other coronavirus sequences	COVID-19, other types of coronaviruses	98.4%
Ozturk et al. (2020)	DarkCovidNet	COVID-19, no-findings, pneumonia	COVID-19, healthy, pneumonia	87.02%
Alakus and Turkoglu, 2020	ANN CNN CNNLSTM CNNRNN LSTM RNN	COVID-19	laboratory findings	92.3%

(*Continued*)

TABLE 6.1 (Continued)

Study	Method Used	Dataset	Image Classes	Performance Measure (Accuracy%)
Asnaoui and Chawki (2020)	VGG16, VGG19, DenseNet201, Inception ResNet V2, Inception V3, Resnet50, MobileNet V2	1,583 normal X-ray and CT images, 231 COVID-19 images, 2,780 images of bacterial pneumonia, 1,493 of coronavirus images	COVID-19, other viral pneumonias	92.18%
Jain et al. (2020)	VGG16 ResNet18 DenceNet121 ResNet101	X-ray images	COVID-19, bacterial pneumonia, viral pneumonia, healthy	98.93%
Brunese et al. (2020)	VGG16	6,523 X-ray images	COVID-19, generic pneumonia	99%
Basu et al. (2020)	Domain Extension Transfer Learning	X-ray images	COVID-19 normal, pneumonia, other diseases	95.3%
Ahsan et al. (2020)	MLP-CNN	X-ray images	COVID-19, non COVID-19	96.3%
Narin et al. (2020)	ResNet50, ResNet101, ResNet152, InceptionV3, Inception, ResNetV2	X-ray images	341 COVID-19, 2,772 bacterial pneumonias, 1,493 viral pneumonias, 2,800 healthy	98%
Xu et al., 2020	multiple CNN deep learning technique	CT scans	219 COVID-19, 224 influenza, 175 healthy	86.7%
Alazab et al. (2020)	CNN	X-ray images	COVID-19, non COVID-19	90–95%

COVID-19 have gained momentum, as neither RT-PCR nor the accuracy of nucleic acid test tests alone are sufficient to detect and diagnose COVID-19 disease status. It is well known from (Narin et al. 2020) the detection of COVID-19 symptoms that CT scans or X-rays diagnosis methods applied to the lower regions of the lungs have a higher accuracy when compared to the RT-PCR test and improve the speed and accuracy of diagnosis.

One of the major limitations of studies on COVID-19 detection and diagnosis with the help of deep learning methods applied to X-ray images and CT scans is seen as the size of the data used during the creation of models. Another important limitation is that further studies must be performed with laboratory findings from other databases or other regions to test the validation of the model. Moreover, a significant point is the fact that the effect of different stages of the disease on the predictive performance of the models needs to be investigated further and then the model should be reviewed accordingly. Advanced deep learning algorithms can identify those most susceptible to COVID-19 disease due to their genetic and physiological characteristics (Alimadadi et al. 2020).

Now we state the challenges and problems related to the application of deep learning methods to X-ray images and CT scans to overcome the COVID-19 pandemic as follows:

(1) The availability of high-quality images and large data sets associated with COVID-19 disease give rise to a huge challenge for patient data privacy (Bhattacharya et al., 2021).

(2) If study participants do not improve or die within the specified period, excluding their data from the analysis performed with the same data set will have a negative effect on the performance of the proposed models (Wynants et al. 2020).

(3) The variability experienced in the COVID-19 test process of various hospitals around the world is an important problem that causes data features not to be considered properly. (Bhattacharya et al., 2021).

(4) Since normal Pneumonia and COVID-19 Pneumonia symptoms are very similar, defining a suitable DL method to diagnose COVID-19 with remarkable accuracy is difficult. (Bhattacharya et al., 2021).

In addition, time-consuming diagnostic procedures performed by medical doctors using CT scans and X-ray images are a major challenge. As seen from the literature, many disease prediction models are being developed to overcome this difficulty by using artificial intelligence-based diagnostic methods.

6.6 CONCLUSION

Deep learning technologies compared to SARS-CoV in 2003 and MERS-CoV in 2012 have been successfully used in almost every area of the COVID-19 outbreak such as diagnosis and treatment, prediction, tracking of virus spread, and drug discovery. In this chapter, we reviewed deep learning methods predicting COVID-19 cases. We investigated these studies in two aspects: studies using genomic data and radiology image data. The most commonly used techniques for diagnosis of COVID-19 disease are RT-PCR testing and medical imaging (chest X-ray and CT scan). CT scans or X-ray diagnosis methods have higher accuracy when compared to the RT-PCR test. Moreover, they provide speed and accuracy of diagnosis. However, when considering the high mutation rate of SARS-CoV-2, using genomic sequences is extremely beneficial when tracking coronavirus genes, which may change frequentlyas the disease spreads from one person to another. We discussed advantages anddrawbacks of the methods in detail. Finally, we gave the main challenges and potential direction to fight against this pandemic.

REFERENCES

Ahsan, M.M., Alam, E.T., Trafalis, T., and Huebner, P. 2020. Deep MLP-CNN model using mixed-Datato distinguish between COVID-19 and non-COVID-19 patients. *Symmetry* 12(9):1526.

Ai, T., Yang, Z., Hou, H., Zhan, C., Chen, C., Lv, W., Tao, Q., Sun, Z., and Xia, L. 2020. Correlation of chest CT and RT-PCR testing for coronavirus disease 2019 (COVID-19) in China: A report of 1014 cases. *Radiology* 296:E32-E40.

Alakus, T.B., and Turkoglu, I. 2020. Comparison of deep learning approaches topredict COVID-19 infection. *Chaos, Solitons& Fractals* 140:110120.

Alazab, M., Awajan, A., Mesleh, A., Abraham, A., Jatana, V., and Alhyari, S. 2020. COVID-19 prediction and detection using deep learning. *International Journal ofComputer Information Systems and Industrial Management Applications* 12:168–181.

Alimadadi, A., Aryal, S., Manandhar, I., Munroe, P. B., Joe, B., and Cheng, X. 2020. Artificial intelligence and machine learning to fight COVID-19. *PhysiologicalGenomics* 52(4):200–202. PMID: 32216577.

Anika, S., Monika, P., Andre, C., Jamie, L. B., Vanessa, S., Joanna, E., Shamez, L., Maria, Z., and Robin, G. 2020. Duration of infectiousness and correlation with RT-PCR cycle threshold values in cases of COVID-19. *Eurosurveillance* 25(32): 2001483.

Annarumma, M., Withey, S. J., Bakewell, R. J., Pesce, E., Goh, V., and Montana, G. 2019. Automated triaging of adult chest radiographs with deep artificial neural networks. *Radiology* 291(1):196–202.

Anwar, S.Y., Majid, M., Qayyum, A., Awais, M., Alnowami, M., and Khan, M.K. 2018. Medical image analysis using convolutional neural networks: A review. *Journal of Medical Systems* 42:226.

Apostolopoulos, I.D., and Bessiana, T. 2020. COVID-19: Automatic detection from X-ray images utilizing transfer learning with convolutional neural networks. *Physical and Engineering Sciences in Medicine* 43(2):635–640.

Arabi, Y.M., Murthy, S., and Webb, S. 2020. COVID-19: A novel coronavirus and a novel challenge for critical care. *Intensive Care Med* 46:833–836.

Arora, P., Kumar, H., and Panigrah, B.K. 2020. Prediction and analysis of COVID-19 positive cases using deep learning models: A descriptive case study of India. *Chaos, Solitons& Fractals* 139:110017.

Arslan, H., and Arslan, H. 2021. A new COVID-19 detection method from human genome sequences using CpG island features and KNN classifier. *Engineering Science and Technology, an International Journal.* https://doi.org/10.1016/j.jestch.2020.12.026.

Asnaoui, K.E., and Chawki, Y. 2020. Using X-ray images and deep learning for automated detection of coronavirus disease. *Journal of BiomolecularStructure and Dynamics* 22:1–12. PMID: 32397844.

Basu, S., Mitra, S., and Saha, N. 2020. Deep learning for screening COVID-19 using chest X-ray images. https://arxiv.org/pdf/2004.10507.pdf.

Bayer, J., Wierstra, D., Togelius, J., and Schmidhuber, J. 2009. *Evolving memory cell structures for sequence learning. International conference on artificial neural networks,* pages 755–764.

Bernal, J., Kushibar, K., Asfaw, D. S., Valverde, S., Oliver, A., Marti, R., and Llado, X. 2019. Deep convolutional neural networks for brain image analysis on magnetic resonance imaging: a review. *Artificial Intelligence in Medicine* 95:64–81.

Bhattacharya, S., Reddy Maddikunta, P.K., Pham, Q.-V., Gadekallu, T.R., Krishnan, S.R., Chowdhary, C.L., Alazab, M., and Jalil Piran, M. 2021. Deep learning and medical image processing for coronavirus (COVID-19) pandemic: A survey. *Sustainable Cities and Society* 65:102589.

Brunese, L., Mercaldo, F., Reginelli, A., and Santone, A. 2020. Learning for pulmonary disease and coronavirus COVID-19 detection from X-rays. *Computer Methods and Programs in Biomedicine* 196:105608.

Chen, G., Li, K., Zhang, Z., Li, K., and Yu, P.S. 2020. A Survey on Applications of Artificial Intelligence in Fighting Against COVID-19. https://arxiv.org/pdf/2007.02202.pdf.

Chimmula, V.K.R., and Zhang, L. 2020. Time series forecasting of COVID-19 transmission in Canada using LSMT networks. *Chaos, Solitons &Fractals* 135:109864.

Cho, K., Merrienboer, B., Bahdanau, D., and Bengio, Y. 2014. *On the properties of neural machine translation: Encoder-decoder approaches*. In *Proceedings of SSST-8, Eighth Workshop on Syntax, Semantics and Structure in Statistical Translation*, pages 103–111. Association for Computational Linguistics.

Dauphin, Y.N., de Vries, H., and Bengio, Y. 2015. Equilibrated adaptive learning rates for non-convex optimization. https://arxiv.org/pdf/1502.04390.pdf.

Eman, H.A., and Mohamed, M. 2016. *House price estimation from visual and textual features*. In *Proceedings of the 8th International Joint Conference onComputational Intelligence - Volume 2: NCTA, (IJCCI 2016)*, pages 62–68. INSTICC, SciTePress.

Erman, J.L. 1990. Finding structure in time. *Cognitive Science* 14(2):179–211.

Fu, M., Yi, S.L., Zeng, Y., Ye, F., Li, Y., Dong, X., Ren, Y.D., Luo, L., Pan, J.S., and Zhang, Q. 2020. Deep learning-based recognizing COVID-19 and othercommon infectious diseases of the lung by chest CT scan images. https://www.medrxiv.org/content/10.1101/2020.03.28.20046045v1.full.pdf

Gers, F., Schmidhuber, J., and Cummins, F. 2000. Learning to forget: continual prediction with LSTM. *NeuralComputation* 12(10):2451–2471.

Ghoshal, B., and Tucker, A. 2020. Estimating uncertainty and interpretability in deep learning for coronavirus (COVID-19) detection. https://arxiv.org/pdf/2003.10769.pdf

Jain, G., Mittal, D., Thakur, D., and Mittal, M. K. 2020. A deep learning approach to detect COVID-19 coronavirus with X-ray images. *Biocybernetics and Biomedical Engineering* 40(4):1391–1405.

He, K., Zhang, X., Ren, S., and Sun, J. 2016. *Deep residual learning for image recognition*. In *2016 IEEE Conference on Computer Vision and Pattern Recognition (CVPR)*, pages 770–778.

Hemdan, E.E.D., Shouman, M.A., and Karar, M.E. 2020. Covidx-net: A framework of deep learning classifiers to diagnose COVID-19 in X-ray images. https://arxiv.org/ftp/arxiv/papers/2003/2003.11055.pdf

Hochreiter, S., and Schmidhuber, J. 1997. Long short-term memory. *Neural Computation* 9(8):1735–1780.

Holshue, M.L., DeBolt, C., Lindquist, S., Lofy, K. H., Wiesman, J., Bruce, H., Spitters, C., Ericson, K., Wilkerson, S., Tural, A., Diaz, G., Cohn, A., Fox, L., Patel, A., Gerber, S.I., Kim, L., Tong, S., Lu, X., Lindstrom, S., Pallansch, M.A., Weldon, W.C., Biggs, H.M., Uyeki, T.M., and Pillai, S.K. 2020. First case of 2019 novel coronavirus in the united states. New England *Journal of Medicine* 382(10):929–936.

Houdt, G.V., Mosquera, C., and Napoles, G. 2020. A review on the long short-term memory model. *Artificial Intelligence Review* 53:5929–5955.

Huang, C.J., Chen, Y.H., Ma, Y., and Kuo, P.H. 2020. Multiple-input deep convolutional neural network model for COVID-19 forecasting in China. https://www.medrxiv.org/content/10.1101/2020.03.23.20041608v1.full.pdf.

Imagenet. 2016. http://www.image-net.org/. Accessed: 2020-10-10.

Jiang, F., Deng, L., Zhang, L., Cai, Y., Cheung, C.W., and Xia, Z. 2020. Review of the clinical characteristics of coronavirus disease 2019 (COVID-19). *Journal of General Internal Medicine* 35:1545–1549.

Kingma, D.P., and Ba, J. 2017. Adam: A method for stochastic optimization. https://arxiv.org/pdf/1412.6980.pdf.

Krizhevsky, A., Sutskever, I., and Hinton, G.E. 2012. *Imagenet classification with deep convolutional neural networks.* In *Proceedings of the 25th International Conference on Neural Information Processing Systems - Volume 1, NIPS'12*, pages 1097–1105, Red Hook, NY, USA. Curran Associates Inc.

LeCun, Y., Boser, B., Denker, J., Henderson, D., Howard, R., Hubbard, W., and Jackel, L. 1989. Backpropagation applied to handwritten zip code recognition. *Neural computation* 1:541–551.

Liu, Y., Chen, X., Wang, Z., Wang, Z.J., Ward, R.K., and Wang, X. 2018. Deep learning for pixel-level image fusion: Recent advances and future prospects. *Information Fusion* 42:158–173.

Michael, J., and Wei, T.Y. 2020. Laboratory diagnosis of emerging human coronavirus infections the state of the art. *Emerging Microbes & Infections* 9(1):747–756.

Naeem, M.S., Mabrouk, M.S., Marzouk, S.Y., and Eldosoky, M.A. 2020. A diagnostic genomic signal processing (GSP)-based system for automatic feature analysis and detection of COVID-19. *Briefings in Bioinformatics* 2020 August 14: 1197–1205. bbaa170.

Narin, A., Kaya, C., and Pamuk, Z. 2020. Automatic detection of coronavirus disease (COVID-19) using X-ray images and deep convolutional neural networks. https://arxiv.org/ftp/arxiv/papers/2003/2003.10849.pdf.

Ozturk, T., Talo, M., Yildirim, A., Baloglu, U., Yildirim, O., and Acharya, U.R. 2020. Automated detection of COVID-19 cases using deep neural networks with X-ray images. *Computers in Biology and Medicine* 121:103792.

Rachna, C. 2017. *Difference between X-ray and CT scan.* https://biodifferences.com/difference-between-x-ray-and-ct-scan.html. Accessed: 2020.11.01

Randhawa, G., Hill, K., and Kari, L. 2019. ML-DSP: Machine learning with digital signal processing for ultrafast, accurate, and scalable genome classification at all taxonomic levels. *BMC Genomics* 20(267): 267–287.

Randhawa, S.G., Soltysiak, M.P.M., Roz, H.E., Souza, C.P.E., Hill, K.A., and Kari, L. 2020. Machine learning using intrinsic genomic signatures for rapid classification of novel pathogens: COVID-19 case study. *PLoS ONE* 15(4):e0232391.

Schuster, M., and Paliwal, K.K. 1997. Bidirectional recurrent neural networks. *Artificial Intelligence Review* 45(11):2673–2681.

Shelhamer, E., Long, J., and Darrell, T. 2017. Fully convolutional networks for semantic segmentation. *IEEE Transactions on Pattern Analysis and Machine Intelligence* 39(4):640–651.

Shi, F., Wang, J., Shi, J., Wu, Z., Wang, Q., Tang, Z., He, K., Shi, Y., and Shen, D. 2020. Review of artificial intelligence techniques in imaging data acquisition, segmentation and diagnosis for COVID-19. *IEEE Reviews in Biomedical Engineering*. doi: 10.1109/RBME.2020.2987975.

Shi, X., Chen, Z., Wang, H., Yeung, D. Y., Wong, W. K., and Woo, W.C. 2015. *Convolutional LSTM network: A machine learning approach for precipitation nowcasting.* In *Proceedings of the 28th International Conference on Neural Information Processing Systems - Volume 1, NIPS'15*, pages 802–810, Cambridge, MA, USA, MIT Press.

Simonyan, K., and Zisserman, A. 2014. Very deep convolutional networks for large-scale image recognition. https://arxiv.org/pdf/1409.1556.pdf.

Solis-Reyes, S., Avino, M., Poon, A., and Kari, L. 2020. An open-source k-mer based machine learning tool for fast and accurate sub-typing of HIV-1 genomes. *PLoS ONE* 13(11): 0206409.

Su, Y., and Kuo, C.C.J. 2019. On extended long short-term memory and dependent bidirectional recurrent neural network. *Neurocomputing* 356:151–161.

Suzuki, K. 2017. Overview of deep learning in medical imaging. *Radiological Physic and Technology* 10(3):257–273.

Ting, D.S.W., Carin, L., Dzau, V., and Wong, T.Y. 2020. Digital technology and COVID-19. *Nature Medicine* 26:459–461.

Wang, S., Kang, B., Ma, J., Zeng, X., Xiao, M., Guo, J., Cai, M., Yang, J., Li, Y., Meng, X., and Xu, B. 2020. *A deep learning algorithm using CT images to screen for corona virus disease (COVID-19)*. https://www.medrxiv.org/content/10.1101/2020.02.14.200230 28v5.full.pdf

Wolfel, R., Corman, V. M., Guggemos, W., Seilmaier, M.,Zange, S., Müller, M. A., Niemeyer, D., Jones, T. C.,Vollmar, P., Rothe, C., Hoelscher, M., Bleicker, T., Brünink, S., Schneider, J., Ehmann, R.,Zwirglmaier, K., Drosten, C., and Wendtner, C. 2020. Virological assessment of hospitlized patients with COVID-2019. *Nature* 581:465–469.

Wynants, L., Calster, B.V., Collins, G.S., Riley, R.D., Heinze, G., Schuit, E., Bonten, M.M.J., Dahly, D.L., Damen, J.A.A., Debray, T.P.A., de Jong, V.M.T., De Vos, M., Dhiman, P., Haller, M.C., Harhay, M.O., Henckaerts, L., Heus, P., Kreuzberger, N., Lohmann, A., Luijken, K., Ma, J., Martin, G.P., Andaur Navarro, C.L., Reitsma, J.B., Sergeant, J.C., Shi, C., Skoetz, N., Smits, L.J.M., Snell, K.I.E., Sperrin, M., Spijker, R., Steyerberg, E.W., Takada, T., Tzoulaki, I., van Kuijk, S.M.J., van Royen, F.S., Verbakel, J.Y., Wallisch, C., Wilkinson, J., Wolff, R., Hooft, L., Moons, K.G.M., and van Smeden, M. 2020. Prediction models for diagnosis and prognosis of COVID-19: systematic review and critical appraisal. *BMJ* 369:m1328.

Xu, X., Jiang, X., Ma, C., Du, P., Li, X., Lv, S., Yu, L., Ni, Q., Chen, Y., Su, J., Lang, G., Li, Y., Zhao, H., Liu, J., Xu, K., Ruan, L., Sheng, J., Qiu, Y., Wu, W., Liang, T., and Li, L. 2020. A deep learning system to screen novel coronavirus disease 2019 pneumonia. *Engineering* 6(10):1122–1129.

Zeiler, M.D., and Fergus, R. 2014. Visualizing and understanding convolutional networks. In Fleet, D., Pajdla, T., Schiele, B., and Tuytelaars, T., editors, *Computer Vision* ECCV 2014, pages 818–833, Cham:Springer International Publishing.

Zhang, C.,Liao, Q., Rakhlin, A., Miranda, B., Golowich, N., and Poggio, T. 2018. Theory of Deep Learning IIb: OptimizationProperties of SGD. https://arxiv.org/pdf/1801.02254. pdf.

Zheng, S., Jayasumana, S., Romera-Paredes, B., Vineet, V., Su, Z., Du, D., Huang, C., and Torr, P.H. S. 2015. *Conditional random fields as recurrent neural networks*. In *2015 IEEE International Conference on Computer Vision (ICCV)*, pages 1529–1537.

7 Applications of Lifetime Modeling with Competing Risks in Biomedical Sciences

N. Chandra and H. Rehman
Pondicherry University, India

CONTENTS

7.1 INTRODUCTION

The classical lifetime data analysis measures time interval from some time point to until the occurrence of primary event of interest. For examples, death of the individual due to certain disease, failure of an equipment, etc., hence their life length is observed, known as complete lifetime data. If the event of interest has not occurred up to follow-up, then the survival time is not known exactly for some patients. Therefore, this phenomenon is known as censoring. For example, after undergoing

DOI: 10.1201/9781003161233-7

breast cancer treatment, a person is monitored for four years and does not experience a recurrence during that time span. However, as the duration of tumour recurrence remains unknown, this patient provides useful information on the likelihood that recurrence will not occur until four years later. Therefore, in this case, the corresponding density function, survival function, cumulative distribution function and hazard function are the quantities that describe the distribution of the occurrence of an event with regard to time. Statistical methods for estimating these quantities includes parametric, nonparametric and semiparametric are well established.

In most of the studies an individual or subject at risk of p mutually exclusive type of events or failure causes, where occurrence of one type of failure alter the chance of the occurrence of other types of failure. Thus such types of failure (events) are refer as competing risks. For example, primary biliary cirrhosis is a chronic liver disease in which individual may receive the transplant and experience the death in waiting queue. Similarly, a breast cancer patient can experience local relapse and death. Therefore, the standard survival models and methods are inappropriate in that situation. After World War II, the various enormous statistical methods have been developed for the analysis of lifetime data in the presence of competing causes (failure/death) in the domain of medical as well as engineering studies. For more detail on competing risk modeling one could refer to the text books Kalbfleisch and Prentice (2002) and Beyersmann et al. (2012).

The theory of competing risks has a fascinating history going back to the eighteenth century. In 1760, Daniel Bernoulli applied Halley's approach to demonstrate the advantages of smallpox inoculation. He calculated the increase in Halley's survivor function if smallpox was eliminated as a cause of death. In this way, Bernoulli founded the theory of competing risks. Bernoulli had considered a mathematical method to solve the problem. He wanted to compare the mean duration of life in two differently constituted populations: a real population, who were subject to death from smallpox and from other causes, versus another, hypothetical population, for whom smallpox was not a cause of death; see Makeham (1874) and Chiang (1961) explored some practical application of the theory.

Latent failure time approach was the first choice for analysis of competing risks modeling ((Moeschberger and David (1971), Crowder (2001)). This method is based on a series system, in which it is assumed that time to event of interest and competing risks are independent. Therefore, it is consider that the inference about the distribution of time to an event of interest is obtained in the hypothetical situation in which it is assumed that other events rather than event of interest are absent. The other methods for analyzing the competing risks survival data are based on the estimation of event specific quantities from the observed data. The basic two specific quantities are cause specific hazard (CSH) function and cumulative incidence function (CIF). The approaches to obtained these quantities are CSH functions (Prentice et al., 1978), mixture model approaches (Larson and Dinse, 1985), subdistribution hazard (SDH) functions (Fine and Gray, 1999) and direct parameterization of CIF (Jeong and Fine, 2006).

The latent failure time approach is mathematically convenient for practitioner to analyzing the competing risks data mostly in the engineering sciences. The independence assumption of latent failure times require a detail knowledge of the structure of the system under study, it is only possible to verify this assumption in very special situations. The main problem arises with this approach while dealing with medical applications of competing risks. For detailed reviews in this direction one could refer to Tsiatis (1975) and Kalbfleisch and Prentice (2002). On the other hand, cause specific quantities such as CSH, CIF and SDH can be estimated from the observed data and need not be used to verify dependency between the competing risks.

This chapter reviews some statistical methods for modeling competing risks with their limitation and advantages. The rest of the chapter is organised as follows. In Section 7.2, we discussed some aspects of latent failure time approach. Section 7.3, represents the various bivariate approach of competing risk such as cause specific proportional hazards approach, subdistribution proportional hazards approach, mixture model and direct parameterization method for estimation of CIF. Different machine learning approaches for analyzing the survival data are given in Section 7.5. In Section 7.6, we provide a real life data for the application of competing risks. In the last, in Section 7.7, discussion and concluding remarks are given.

7.2   LATENT FAILURE TIME APPROACH IN COMPETING RISKS

Initially, the problems of competing risks were formulated in terms of latent or potential failure times, and it is the classical approach for modeling the competing risks data. Cox (1959), and Moeschberger and David (1971) described the competing risk model based on the latent/hypothetical failure times. Suppose that $T_1, T_2, \ldots, T_p$ be the p random variables according to p causes of failure. In real life scenario, only the time to the first event denoted as T_c, with

$$T_C = \min\left(T_1, T_2, \ldots, T_p\right) \tag{7.1}$$

can be observed. The index C is the set of p possible causes of failure i.e. $C \in \{1, 2, \ldots, p\}$. Therefore, the multivariate joint survivor function is defined as

$$S\left(t_1, t_2, \ldots, t_p\right) = Pr\left(T_1 > t_1, T_2 > t_2, \ldots, T_p > t_p\right). \tag{7.2}$$

Hence the sub densities can be calculated from the joint survival function of latent failure times as

$$f_j\left(t\right) = \left(-\frac{\partial S\left(t_1, t_2, \ldots, t_p\right)}{\partial t}\right)_{t_1 = t_2 = \ldots = t_p}, \tag{7.3}$$

where $j = 1, 2, \ldots, p$. Marginal survival function for cause j is given by

$$S_j\left(t\right) = S\left(t_1 = 0, \ldots, t_j = t, \ldots, t_p = 0\right). \tag{7.4}$$

The fundamental problem with latent failure time approach is that without independence assumptions, the joint survival function is not identifiable from the observed data (Tsiatis, 1975). The assumption of independence of hypothetical failure times is a strong one, therefore, it can rarely be justified in real life situation. Unfortunately, when only the first event is observed this assumption cannot be tested. Thus this problem refer to as problem of non-identifiability of marginal survival function of latent failure times. For detail on latent failure time approach one could refer to Crowder (2001) and reference therein. To overcome the difficulty of latent failure time approach, researchers developed the hazard based methods using bivariate approach, which are discussed in next section.

7.3 BIVARIATE APPROACH OF COMPETING RISKS

The alternative approach of competing risks analysis is to consider the bivariate random variable (T, C), where T is the continuous random variable for specific event time and C is the discrete random variable denoting the types of the event. Based on this approach recently hazard-based methods get considerable attention in statistical literature. A detailed discussion towards the different methods of competing risks is available in Kalbfleisch and Prentice (2002) and Beyersmann et al. (2012). The following subsections described the most important and commonly used concepts of competing risks modeling.

7.3.1 CUMULATIVE INCIDENCE FUNCTION

In the lifetime modeling, when the competing risks are present, caution is needed in estimating the probability of the occurrence of the event of interest. The Kaplan–Meier method is the most common and relatively easy to use in survival analysis. The availability of Kaplan–Meier method in statistical packages makes its use appealing. However, if the competing risks are presents then the complement of Kaplan–Meier survival function over estimate the cumulative probability of failure, because it consider the competing causes as censored event while estimating the survival function for event of interest. Therefore, considering the events of the competing causes as censored observations will lead to a bias in the Kaplan–Meier estimate (Pintilie, 2006).

According to Kalbfleisch and Prentice (2002) CIF is the cumulative probability of failure due to particular cause j in the presence of all other risks and can be defined as

$$F_j(t) = P(T \le t, C = j). \qquad (7.5)$$

In statistical literature CIF is also known as crude event probability, improper function and subdistribution function. CIF is called an improper function in the sense that asymptotes of CIFs are less than 1, as it does not converge to 1 for t going to infinity, but to the overall probability for an specific event (π_j) of type j

$$\lim_{t \to \infty} F_j(t) = P(C = j) = \pi_j. \qquad (7.6)$$

with the condition

$$\sum_{j=1}^{p} \pi_j = 1.$$ (7.7)

The overall distribution function is the probability that an event of any type occurs at or before time t is given by

$$F(t) = \sum_{j=1}^{p} F_j(t).$$ (7.8)

To estimate the CIF, various statistical methods have been developed which are discussed in next subsections.

7.3.2 CAUSE SPECIFIC HAZARD FUNCTION

In classical lifetime data analysis, hazard rate play a vital role, similarly the CSH play an important role in competing risks modeling. According to Prentice et al. (1978), without checking the independence assumption, the competing risks can be measured using hazard-based methods and all measures can be derived from observable data. CSH rate is the basic estimable quantity of competing risks framework, which is defined as the probability of failure of a subject from a particular type of cause j in small interval t to $t + \Delta t$ given that none of the cause occur before time t.

In other words, the CSH function is the instantaneous failure rate for a given cause in subject who are currently event free. The CSH rate mathematically defined as

$$\lambda_j(t) = \lim_{\Delta t \to 0} \frac{P(t \leq T < t + \Delta t, C = j | T \geq t)}{\Delta t} = \frac{f_j(t)}{S(t)},$$ (7.9)

where $j = 1, 2, ..., p$ is denote the possible number of failure causes. We are assuming that causes are mutually exclusive end points, then the over all hazard for all p causes at time t is the sum of all CSH rate from any cause at time t is

$$\lambda(t) = \sum_{j=1}^{p} \lambda_j(t)$$ (7.10)

As in classical survival analysis, for the competing risks the cumulative CSH rate for a cause j at time t is the integral of the CSH function over 0 to t, i.e. $\Lambda_j(t) = \int_0^t \lambda_j(u) du$.
Then over all survivor function which is the probability of being free from any risk, depends upon the total cumulative hazard, which is nothing but the sum of all cumulative CSH functions

$$S(t) = \exp\left(-\Lambda(t)\right) = \exp\left(-\sum_{j=1}^{p}\Lambda_j(t)\right). \tag{7.11}$$

where $\Lambda(t) = \sum_{j=1}^{p}\Lambda_j(t)$. From the standard relationship between distribution function, survivor function and hazard function, the CIF for cause j in terms of CSH function is defined as

$$F_j(t) = P(T \le t, C = j) = \int_0^t \lambda_j(u)S(u)du = \int_0^t \lambda_j(u)\exp\left(-\sum_{j=1}^{p}\Lambda_j(u)\right)du. \tag{7.12}$$

If the explanatory variables are continuous or the simultaneous effect of several covariates on cause-specific failure is of interest, then the regression model can be developed by using Cox proportional hazards approach (Cox, 1972) with a flexible mis-specified baseline CSH rate $\lambda_{0j}(t)$ as follows

$$\lambda_j(t|x) = \lambda_{0j}(t)\exp\left(\beta_j^{\mathrm{T}}x\right), \tag{7.13}$$

where x is the covariate vector and β_j represents the covariate effects on cause j. The proportionality hazard assumption can be checked by Andersen plot (Andersen, 1982). According to Beyersmann et al. (2012) the CSH function completely determine the competing risks process, so CIFs can be estimated from CSH regression models for all types of event. Therefore, CIF based on CSH function with covariate x is defined as

$$F_j(t|x) = \int_0^t \lambda_j(u|x)\exp\left(-\sum_{j=1}^{p}\Lambda_{0j}(u)\exp\left(\beta_j^{\mathrm{T}}x\right)\right)du, \tag{7.14}$$

where $\Lambda_{0j}(t)$ is the baseline cumulative CSH function.

Modeling of competing risks data through CSH functions can be conducted in standard statistical software packages via classical Cox regression treating failures from the cause of interest as events of interest and failures from other risks as censored observations. The analysis is completely straightforward and standard, but the interpretation requires caution. The interpretation of CIF in terms of CSH is little awkward in group comparison, as can be seen from (7.10), the CIF for a cause j depends on the all p types of CSH. This issue is raised in various articles; see (Putter et al. (2007), Allignol et al. (2011), Haller et al. (2013)). As a consequence of that fact, CSH for an event of type j for one strata does not mean to translate a higher cumulative probability of that event. Therefore, the CSH feature has no one-to-one relationship with CIF, so the model of subdistribution hazards is built for this reason, which is discussed in the next subsection.

7.3.3 Subdistribution Hazard Function

The applications CIF, or subdistribution function, in competing risks modeling is well recognized in epidemiology, demography and survival analysis. One can see from Equation (7.14), the CSH function does not directly linked with CIF due the sum of all cumulative CSH appearing in the exponential term. To overcome the problem of direct relationship between CSH function and CIF, Gray (1988) introduced a model called the SDH model for comparing the CIF in k different groups. The SDH function is defined as the instantaneous rate of failure of the given cause in subject to who have not yet experienced of a cause of that type. Mathematically, it can be written as

$$\gamma_j(t) = \lim_{\Delta t \to 0} \frac{P\left(t \leq T \leq t + \Delta t, C = j, T \geq t \cup \{T < t, C \neq j\}\right)}{\Delta t}$$

$$\gamma_j(t) = \frac{\frac{d}{dt} F_j(t)}{1 - F_j(t)} = \frac{-d \log(1 - F_j(t))}{dt} \tag{7.15}$$

The relationship between the CIF and the SDH turns out in the following form

$$F_j(t) = 1 - \exp\left(-\Gamma_j(t)\right), \tag{7.16}$$

where $\Gamma_j(t)$ is the cumulative SDH, defined by

$$\Gamma_j(t) = \int_0^t \gamma_j(u) \, du. \tag{7.17}$$

When events are occurring in discrete time, and censoring is absent, the SDH at time t_i can be estimated as

$$\hat{\gamma}_j(t_i) = \frac{d_{ji}}{n_i^*} \tag{7.18}$$

where d_{ji} is the number of failure of type j at time t_i and n_i^* is the corresponding risk set including all individuals who are censored at time t_i and all individuals that failed before t_i from a cause other than j. Many authors have raised the issue of risk set associated with SDH model (Haller et al., 2013).

In order to avoid the difficulty arising with CSH in the interpretation of the effect of regression coefficient on CIF, Fine and Gray (1999) proposed the proportional SDH model for event of interest or primary event of failure. The SDH model for primary cause is popular, because of its direct relationship with CIF for same cause, which is not found in CSH function. Due the wide popularity and importance of Cox proportional hazards model, it can be employed with SDH model as

$$\gamma_j(t|x) = \gamma_{0j}(t) \exp\left(\beta_j^T x\right), \tag{7.19}$$

where $\gamma_{0j}(t)$ is the baseline SDH function. Therefore, the SDH is plugged directly to the CIF in a way as in classical survival analysis with one event of interest as

$$F_j(t|\mathbf{x}) = 1 - \exp\left(\Gamma_j(t|\mathbf{x})\right) = 1 - \exp\left(\int_0^t \gamma_j(t|\mathbf{x})\right) \tag{7.20}$$

Hence the CIF for the event of interest can be estimated directly from the regression coefficients and baseline SDH obtained by a Fine and Gray model without specific relationship of the covariate effects on competing events. This model can not be used simultaneously for two competing causes. Therefore, to estimate CIF for competing event we need to run this model twice. In general, the proportionality assumption cannot hold true for separate SDH regression models for different types of events (Beyersmann et al., 2012). However, estimation of regression coefficients follows the partial likelihood approach used in a standard Cox model. The SDH and CSH methods are based on hazard function, the alternative of both is the mixture model approach which is discussed in next subsection.

7.3.4 MIXTURE MODEL APPROACH

Another approach for modeling of failure time data in the presence of competing risks is published by Larson and Dinse (1985). They proposed a model to express the joint distribution of event times and type of events as the product of conditional distribution of event times given the type of events and marginal distribution of event type. This can be represented by

$$P(T,C) = P(T|C)P(C), \tag{7.21}$$

where C denoting discrete random variable for the event type and T is a continuous non-negative random variable for the time to failure. The marginal event time probability $P(C)$ is denoted as p_j. In case of two possible causes $p_1 = 1 - p_2$.

The CIF for event type j can be derived from a mixture model as

$$F_j(t) = p_j \bar{F}_j(t). \tag{7.22}$$

where $\bar{F}_j(t)$ denotes the cumulative distribution for the conditional event time distribution given an event of type j and it is a proper distribution function with the conditions $\lim_{t \to \infty} \bar{F}_j(t) = 1$. Therefore, the overall survival function denoting an individuals probability of being free from any event up to time t, given by

$$S(t) = \sum_{j=1}^{p} p_j \bar{S}_j(t) \tag{7.23}$$

Larson and Dinse (1985) developed the regression model for mixture model approach to assess the effect of explanatory variables on the event type probabilities through a multinomial logistic regression model

$$P(C = j|x) = \frac{\exp(\theta_j + \beta_j^T x)}{\sum_{j=1}^{p} \exp(\theta_j + \beta_j^T x)}, \tag{7.24}$$

where θ_j is the scalar constant and β_j is column vector of m regression coefficients and $j = 1, 2, \ldots, p$. They also set the condition for uniqueness of the model as $\theta_p = 0$ and $\beta_p = 0$. An appropriate assumption for the conditional distribution of the lifetime for a given type of event has to be made. Different lifetime distributions were proposed in the literature, like piecewise exponential distribution, three-parameter generalized gamma distribution (Lau et al., 2008), etc.

Each conditional failure time distribution is assumed to depend on covariates through a regression model for the associated hazard function (Cox, 1972). To model the survival function for a given type of failure and a given set of covariates can be obtained as

$$\bar{S}_j(t|x) = P(T > t|x, C = j) = \exp\left(-\int_0^t h_{0j}(u)\exp(\beta_j^T x)du\right), \tag{7.25}$$

where h_{0j} denotes baseline hazard function for a individual with all covariates set to zero. Using an exponential model, the baseline hazard function has the following form

$$h_{0j}(t) = \alpha \tag{7.26}$$

which represents the constant hazard for an event of type j.

Formulation of the likelihood function for the mixture model (Haller et al., 2013) may be complicated, which is one of the major drawbacks of this approach. Lack of the availability of mixture model approach in standard statistical software makes its use difficult for parameter estimation or inference. Expectation-maximization algorithm can be used for obtaining maximum-likelihood estimates of the regression coefficients. Bootstrapping methods can be utilized for estimating the standard errors and confidence intervals for the regression coefficients. In addition to these bivariate methods, some modifications to the bivariate approach have recently been developed, which are given in the next section.

7.4 FURTHER GENERALIZATION OF BIVARIATE APPROACH

In this section, we addressed some recent developments in the bivariate approach to competing risks. Such generalization includes direct parameterization of CIF and fully specified SDH model.

7.4.1 DIRECT PARAMETERIZATION OF CUMULATIVE INCIDENCE FUNCTION

Fine and Gray (1999) proposed the semiparametric model for estimating the CIF by extending the Cox proportional hazards model into the competing risks setting. This model will not work for simultaneous modeling of events of interest as well as competing events. An alternative to this model, Jeong and Fine (2006) proposed the direct parameterization of CIF through improper Gompertz distribution (Gompertz, 1825) without covariates. The two parameter improper Gompertz distribution for parametrizing the CIF is given by

$$F_j(t;\Theta_j) = 1 - \exp\left\{\frac{\alpha_j}{\lambda_j}\left(1 - e^{\lambda_j t}\right)\right\}, \tag{7.27}$$

where $\Theta_j = (\alpha_j, \lambda_j), j = 1, 2, \ldots, p$ is the vector of parameters and $-\infty < \lambda_j < \infty, \alpha_j > 0$. An improper function of (7.27) occurs when $-\infty < \lambda_j < 0$ and $\alpha_j > 0$.

The likelihood function using direct parametric form of CIF can be expressed as follows

$$L = \prod_{i=1}^{n}\left\{f_j(t_i,\Theta_j)\right\}^{I(C_i=j)}\left\{1 - \sum_{j=1}^{p}F_j(t_i;\Theta_j)\right\}^{1-\sum_{j=1}^{p}I(C_i=j)}, \tag{7.28}$$

where $f_j(t,\Theta_j) = dF_j(t;\Theta_j)/dt$.

The important feature of survival data with competing risks is the assumption that the subject will eventually experience the event of interest or the competing event. In this setting, the probability of an event of type j never occurring equals $\lim_{t\to\infty}(1 - F_j(t;\Theta_j)) = \exp\left(\frac{\alpha_j}{\lambda_j}\right)$. Therefore, CIFs from all the causes should add up to one as time goes to ∞ such that

$$F_1(\infty;\Theta_1) + F_2(\infty;\Theta_2) + \ldots + F_p(\infty;\Theta_p) = 1. \tag{7.29}$$

The additivity constraint (7.29) will be well explain in the situation when death is one of the competing risks. For example, if we had followed all the patients long enough in a liver transplant clinical trial, we would have found that each patient either had a transplant, or died without a transplant. The inclusion of covariates is straightforward by following the proportional hazards framework as described in Fine and Gray (1999).

7.4.2 FULL SPECIFIED SUBDISTRIBUTION HAZARD MODEL

Ge and Chen (2012) suggested a fully specified sub-distribution model (FSS) as a generalization of the Fine and Gray (1999) model, for estimating the CIF for events of interest as well as competing events. They considered two causes of failure know as cause 1 and cause 2 for event of interest and competing event respectively. The generalization for more than two competing causes is straightforward. They defined a random variable $T_j^* = T_j \times I(C = j) + \infty \times I(C \neq j), j = 1, 2$, and let $T^* = \min\{T_1^*, T_2^*\}$ be the actual time to failure. They considered the CIF of T^* for two causes as

$$F_1(t) = P(T^* \leq t, C = 1) = P(T_1 \leq t, C = 1) \tag{7.30}$$

and

$$F_2(t) = P(T^* \leq t, C = 2) = M_2(t) P(C = 2) = M_2(t) \pi_2 \tag{7.31}$$

where M_2 is the CIF conditional on cause 2, $P(T_2 \leq t \mid C = 2)$ and $\pi_2 = P(T_2 < \infty, C = 2)$ is the total proportion of failure due to cause 2. From (7.30) and (7.31) the correlation structure between T_1 and T_2 can not be observed, instead this can be developed via (7.29).

Suppose that $h_{10}(t)$ is the baseline SDH function with corresponding cumulative SDH $H_{10}(t) = \int_0^t h_{10}(u) du$, and it is an improper function with the condition $\lim_{t \to \infty} H_{10}(t) < \infty$. Similarly, let $h_{20}(t)$ be the conditional baseline hazard function with corresponding baseline cumulative hazard function $H_{20}(t) = \int_0^t h_{20}(u) du$. Ge and Chen (2012) construct a Cox proportional hazards structure with covariate vector $\mathbf{x}$ and regression coefficient β_1 and β_2 as follows

$$h_1(t|\mathbf{x}) = h_{10}(t) \exp(\beta_1^T \mathbf{x})$$

and

$$h_2(t|\mathbf{x}) = h_{20}(t) \exp(\beta_2^T \mathbf{x}).$$

Therefore, for cause 1, CIF is given by

$$F_1(t|\mathbf{x}) = 1 - \exp\{-H_{10}(t) \exp(\beta_1^T \mathbf{x})\} \tag{7.32}$$

and for cause 2, the conditional CIF is given by

$$M_2(t|\mathbf{x}) = 1 - \exp\{-H_{20}(t) \exp(\beta_2^T \mathbf{x})\}. \tag{7.33}$$

The model defined by (7.32) and (7.33) is thus called the FSS model. Under an FSS model, $P(C = 2|\mathbf{x}) = 1 - P(C = 1|\mathbf{x}) = \exp\{-H_{10}(\infty) \exp(\beta_1^T \mathbf{x})\}$.

Assume there are n observations in the study with observed data (T_i, C_i, x_i), $i = 1$, $2, \ldots, n$. The indicator function $I(C_i = j)$, $j = 1, 2$ takes the possible values 0, 1 and 2 corresponding to "censored", died due to "cause 1" and died due to "cause 2" for the ith individual respectively. Therefore, the likelihood function under FSS model is given by

$$
L_{FSS}(\beta_1, \beta_2, h_{10}, h_{20} \,|\, (T_i, C_i, \mathbf{x}_i)
$$

$$
= \prod_{i=1}^{n} \left[h_{10}(t_i) \exp\left(\beta_1^T \mathbf{x}_i\right) \exp\left\{ -H_{10}(t_i) \exp\left(\beta_1^T \mathbf{x}_i\right) \right\} \right]^{I(C_i = 1)}
$$

$$
\times \left[h_{20}(t_i) \exp\left(\beta_2^T \mathbf{x}_i\right) \exp\left\{ -H_{20}(t_i) \exp\left(\beta_2^T \mathbf{x}_i\right) - H_{10}(\infty) \exp\left(\beta_1^T \mathbf{x}_i\right) \right\} \right]^{I(C_i = 2)} \quad (7.34)
$$

$$
\times \left[\exp\left\{ -H_{10}(t_i) \exp\left(\beta_1^T \mathbf{x}_i\right) \right\} - \left(1 - \exp\left\{ H_{20}(t_i) \exp\left(\beta_2^T \mathbf{x}_i\right) \right\} \right) \right.
$$

$$
\times \left(\exp\left\{ -H_{10}(\infty) \exp\left(\beta_1^T \mathbf{x}\right) \right\} \right) \right]^{I(C_i = 0)}.
$$

The FSS model is not only a generalization of SDH model of Fine and Gray (1999) but also provide a novel justification of SDH partial likelihood under certain condition. For more detail one could refer to Ge and Chen (2012). So far we have discussed various statistical methods for modeling survival time with competing risks. In next Section, we discussed some important methods of machine learning for analyzing the time to event data.

7.5   CONCEPT OF MACHINE LEARNING IN SURVIVAL ANALYSIS

Modeling and analysis of survival data using machine learning techniques is the topic of recent interest in lifetime studies. In the above sections we have discussed various models of survival analysis with competing risks. In this section we will highlight various machine learning models for handling survival data. In recent years, with the rise of big data handling processes, there is an increasing interest in the use of machine learning algorithms, especially those for regression tasks. Machine learning found applications in the field of survival analysis, and researchers developed various methods for handling large number of observations with finite dimension space. Machine learning approaches includes various algorithms such as neural network, survival trees, Bayesian methods, support vector machines and advanced machine learning approaches.

The machine learning methods have attracted practitioners due to various advantages over statistical methods in survival analysis. For example, in Cox regression models, we have to consider the tied lifetime observations and used various specific method to deal the problem. Also, the Cox proportional hazards assumption is violated when the covariates are time dependent. However, while analyzing survival data using machine learning approaches these types of problem are not appearing. For more detail on machine learning method application in survival analysis one could refer to Wang et al. (2019) and Zupan et al. (2000).

7.5.1 ARTIFICIAL NEURAL NETWORKS

Artificial neural networks are one of popularly used nonparametric methods of machine learning approaches in the analysis of lifetime data (Mitchell, 1997). The structure of artificial neural networks has been inspired by biological neural systems. In this method, the simple nodes denoted by neurons are connected based on a weighted link to construct a network that simulates a biological neural network. In general, structure of artificial neural networks contains an input layer, hidden layer and output layer (Cheng and Titterington, 1994). Hidden layers are internally connected with hidden neurons and represent the present non-linear relationship between inputs and outputs. The simplest structure of artificial neural networks has one hidden layer and one hidden neuron. All neurons are fully connected with weights. Sreedevi and Sankaran (2013) analyzed the survival data with competing risks through multi-layer perceptron neural network model.

Both the number of hidden layers and the number of neurons in each hidden layer require experiments to optimize. Both hidden layers and output layer take activation functions which are selected according to target variables. Each neuron takes the weighted sum from previous outputs, transfers it through an activation function, and outputs this value to its downstream neuron. Artificial neural networks are used for many fields of pattern classification and pattern recognition: speech recognition and speech generation, location of radar point sources, target recognition and mine detection, recognition of coins of different denominations, in robotics and computer vision, and many more.

In addition to survival analysis, three kinds of neural network methods are employed in literature to solve the survival analysis problems. The survival time of an individual can be predicted directly from the given inputs using neural network methods. Faraggi and Korn (1996) extended the idea of Cox proportional hazards model to the non-linear artificial neural network predictor and suggested to fit the neural network which has a linear output layer and single logistic hidden layer. Mariani et al. (1997) used both neural network as well as standard Cox model to observe the effect of prognostic factor for the recurrence of breast cancer. A brief description of the neural network methods in the context of survival analysis is available in Wang et al. (2019) and the references cited in.

7.5.2 SURVIVAL TREES

The development of survival trees grew from the mid–1980s up to the mid–1990s, where the goal was mainly to extend existing tree methods to the case of survival data with censoring (Bou-Hamad et al., 2011). Survival trees are one form of classification and regression trees which are tailored to handle censored data. Tree-based algorithms are useful due to their reasonable, interpretable structure as well as their ability to detect complex interactions between covariates. However, traditional tree algorithms require complete observations of the dependent variable in training data, making them unsuitable for censored data.

The basic intuition behind the tree-based method is to recursively partition the data based on a particular splitting criterion, and the objects that are same kind to each other based on the event of interest will be placed in the same node. Recursive partitioning techniques are a popular alternative to parametric models. When applied to survival data, survival tree algorithms partition the covariate space into smaller and smaller regions (nodes) containing observations with homogeneous survival outcomes. The survival distribution in the final partitions (leaves) can be analyzed using a variety of statistical techniques, such as Kaplan-Meier curve estimates.

The main difference between a standard decision tree and the survival tree is in the selection of splitting criterion. The decision tree method performs recursive partitioning on the data by setting a threshold for each feature, however, it can neither consider the interactions between the features nor the censored information in the model (Safavian and Landgrebe, 1991). Most recursive partitioning algorithms generate trees in a top-down, greedy manner, which means that each split is selected in isolation without considering its effect on subsequent splits in the tree.

The splitting criteria used for survival trees can be grouped into two categories: maximizing between-node heterogeneity and minimizing within-node homogeneity. The first class of approaches minimizes the loss function using the within-node homogeneity criterion. Bou-Hamad et al. (2011) discussed important aspect of building a survival tree is the selection of the final tree. The first one is a backward method which builds a large tree and then selects an appropriate sub-tree by pruning some of its branches. The second one is a forward method which uses a built-in stopping rule to decide when to stop splitting a node further. Relatively few survival tree algorithms have been implemented in publicly available, well-documented software. Two user-friendly options are available in **R** packages: **rpart** and **party**.

7.5.3 Bayesian Methods of Machine Learning

Bayesian machine learning is a paradigm for constructing statistical models based on Bayes' theorem. Bayes' theorem is one of the popular fundamental concept in probability theory and mathematical statistics; it provides a link between the posterior probability and the prior probability through observed information, so that one can see the changes in probability values before and after accounting for a certain event. According to Friedman et al. (1997), based on Bayes' theorem there are two classifier, namely, Naive Bayes and Bayesian network. These methods, which provide the probability of the event of interests as their outputs, are commonly studied in the context of clinical prediction. Naive Bayes, a well-known stochastic method in machine learning, is one of the simplest yet effective prediction algorithms. Bayesian networks are directed acyclic graphs that allow efficient and effective representation of the joint probability distribution over a set of random variables. The detail discussion on applications of Bayesian machine learning methods in survival analysis are available in Wang et al. (2019) and references therein.

The main objective of this section is to highlight and promote the machine learning approach for analysing the survival data. Here we have discussed three basic methods of machine learning but it has a vast literature. Support vector machine learning method is a very popular supervised learning approach which used mostly for classification and can also be modified for regression analysis and easily employed in survival analysis. In last few years, more advanced machine learning models have been developed to deal with censored survival data. Ensemble learning (bagging survival tree, random survival forest, boosting), active learning, transfer learning and multitask learning are recent developed methods of learning in the domain of survival analysis.

7.6   APPLICATION TO BREAST CANCER DATA

For illustration purposes, we analyzed a dataset of randomized clinical trial patients with node negative breast cancer. This study was conducted between 1992 and 2000, where a total of 769 women were enrolled with early breast cancer, in which, 383 women were randomly assigned to receive the tamoxifen-alone arm and 386 to receive the combination of radiation and tamoxifen arm. The last follow-up was conducted in the summer of 2002. Only those patients accrued at a single contributor institution are included here: 321 patients in the tamoxifen arm and 320 in the radiation and tamoxifen arm. The median duration of survival time was 5.4 years. An event of interest was local relapse, but some patients experienced axillary relapse, distant relapse, second malignancy of any type, and death. For this study we consider local relapse as a primary event and death without local relapse consider as competing event. The patients who do not experienced local relapse and death considered as right censored observations. This dataset has various covariates such as treatment, age, haemoglobin, etc. We include treatment and age (median age 67 years old) as covariates in the analysis.

First we analyzed this dataset using CSH approach, for this purpose a Cox proportional hazard model was used for each of the two types (death and local relapse) of failure. To observe the effect of covariates we used R statistical software with *coxph* function from the *survival package*.

From Table 7.1 it is observed that treatment has a significant effect on local relapse but not on death. The CSH ratio between the radiation and tamoxifen, and tamoxifen alone groups for death of 1.5645 was observed. This informed that the individual has about 1.5 times higher risk of dying when receiving the radiation and tamoxifen treatment as compared to tamoxifen alone. On the other hand the risk of experiencing the local relapse in radiation and tamoxifen is approximately 10% lower as compared to tamoxifen alone group. Age had more influence on CSH for death with CSH ratio 2.306 compared to 0.6791 for local relapse. The CIFs were estimated for death and local relapse with both the treatment groups and shown in Figure 7.1.

TABLE 7.1
Estimates of Regression Parameters with Hazard Ratio (HR) Using CSH Approach

	$\hat{\beta}$	HR	SE	p-value
Death				
Treatment	0.4476	1.5645	0.3159	0.1566
Age >67	0.8335	2.3060	0.3338	0.0123
Local relapse				
Treatment	−2.2509	0.1053	0.5283	<0.001
Age >67	−0.3870	0.6791	0.3315	0.2430

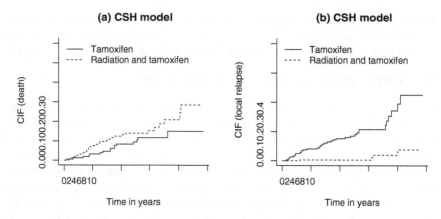

FIGURE 7.1   Estimated CIFs for death and local relapse with treatment as covariate using CSH model (CSH model).

TABLE 7.2
Estimates of Regression Parameters with Hazard Ratio (HR) Using Fine-Gray SDH Approach

	$\hat{\beta}$	HR	SE	p-value
Death				
Treatment	0.5105	1.6661	0.312	0.1
Age >67	0.8326	2.2993	0.3342	0.013
Local relapse				
Treatment	−2.262	0.1041	0.5223	<0.001
Age >67	−0.4048	0.6671	0.3244	0.21

The SDH model described in subsection 3.3 was fit to assess the effect of treatment and age on the SDH rate for death and local relapse. The estimates were obtained using the function *crr* in the R *survival package*, the results are given in Table 7.2. Effects on the SDH can be directly translated to effects on the CIF. For death the SDH ratio of treatment is 1.66 indicates that the patients who received tamoxifen alone treatment are on low risks of dying compared to radiation and tamoxifen. Similarly, the SDH ratio of treatment for local relapse is 0.1041 indicates the patients receiving the tamoxifen alone are approximately 10% of high risk of experience local relapse compared to radiation and tamoxifen. As in CSH approach age has similar effect on death in SDH approach. The estimated CIFs for death and local relapse are shown in Figure 7.2.

7.7   DISCUSSION AND CONCLUSION

In this study we have discussed some basic fundamentals of competing risks modeling, which have their own advantages and disadvantages. The illustration of latent failure time approach is not realistic in real life situations because of the its

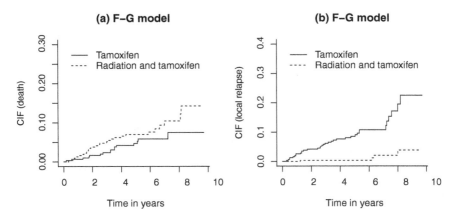

FIGURE 7.2   Estimated CIFs for death and local relapse with treatment as covariate using Fine-Gray model (F-G) model.

independence assumption. CSH functions have the same interpretation as hazard functions in the classical survival analysis, but do not have direct relation with CIF. The SDH functions get some attention, because of the direct relation with CIF, but the risk set related to SDH functions is awkward. We realized that analysis of competing risks data in terms of CSH functions is somehow better than others, because it can directly estimable from the observe data and its availability in various statistical packages. For detailed study of competing risk, follow these references (Haller et al. (2013); Beyersmann et al. (2012); Kalbfleisch and Prentice (2002)).

In the literature it is observed that the CSH and SDH are popular for analyzing competing risks data. Because these methods are easy to apply and availability of these methods in statistical software such as R and SAS makes their use appealing. However, mixture model, direct parameterization methods and FSS modeling have little attention of the practitioner because of lack of availability of statistical packages in R and SAS. So, there are possibilities to explore the use of these methods by developing some statistical packages and make their use easy to practitioners. The availability of recent data in a medical context would make the use of competing risks analysis more popular.

Censoring plays an important role in lifetime modeling, in the presence of competing risks; CSH and SDH methods are analyzed with various types of censoring, such as current status censoring, interval censoring, doubly censoring, etc. So far, mixture model, direct parametrization of CIF and FSS models are considered for right censored competing risks data. Therefore, extension of these models into different types of censoring will be the future research topics. Ge and Chen (2012) considered Bayesian method for the estimation of regression parameters for FSS models. However, mixture model and direct parameterization of CIF have no contribution towards the Bayesian estimation. These are some possibilities of research for the analysis of competing risks through mixture model, direct parameterization method and FSS models have to be explored in future. In this review, we highlighted some fundamental methods of machine learning for survival data analysis with censoring.

But we found that the study of survival data with competing risks does not receive much attention in the machine learning approach. Therefore, it seems that designing machine learning techniques with several forms of failure is a possible opportunity for the researcher to create.

REFERENCES

Allignol, A., Schumacher, M., Wanner, C., Drechsler, C., and Beyersmann, J. (2011). Understanding competing risks: a simulation point of view. *BMC Medical Research Methodology*, 11(1):86.

Andersen, P. K. (1982). Testing goodness of fit of cox's regression and life model. *Biometrics*, 38(1):67–77.

Beyersmann, J., Allignol, A., and Schumacher, M. (2012). *Competing risks and multistate models with R.* Springer Science & Business Media.

Bou-Hamad, I., Larocque, D., Ben-Ameur, H., et al. (2011). A review of survival trees. *Statistics Surveys*, 5:44–71.

Cheng, B., and Titterington, D. M. (1994). Neural networks: a review from a statistical perspective. *Statistical Science*, 21(1):2–30.

Chiang, C. L. (1961). *On the probability of death from specific causes in the presence of competing risks*. In *Proceedings of the fourth Berkeley symposium on mathematical statistics and probability*, volume 4, pages 169–180. Univ of California Press.

Cox, D. R. (1959). The analysis of exponentially distributed life-times with two types of failure. *Journal of the Royal Statistical Society: Series B: Methodological*, 21(2):411–421.

Cox, D. R. (1972). Regression models and life-tables. *Journal of the Royal Statistical Society: Series B: Methodological*, 34(2):187–202.

Crowder, M. J. (2001). *Classical competing risks*. Chapman & Hall/CRC.

Faraggi, D., and Korn, E. L. (1996). Competing risks with frailty models when treatment affects only one failure type. *Biometrika*, 83(2):467–471.

Fine, J. P., and Gray, R. J. (1999). A proportional hazards model for the subdistribution of a competing risk. *Journal of the American Statistical Association*, 94(446):496–509.

Friedman, N., Geiger, D., and Goldszmidt, M. (1997). Bayesian network classifiers. *Machine Learning*, 29(2–3):131–163.

Ge, M., and Chen, M.-H. (2012). Bayesian inference of the fully specified subdistribution model for survival data with competing risks. *Lifetime Data Analysis*, 18(3):339–363.

Gompertz, B. (1825). On the nature of the function expressive of the law of human mortality, and on a new mode of determining the value of life contingencies. *Philosophical Transactions of the Royal Society of London*, (115):513–583.

Gray, R. J. (1988). A class of k-sample tests for comparing the cumulative incidence of a competing risk. *The Annals of Statistics*, 16(3):1141–1154.

Haller, B., Schmidt, G., and Ulm, K. (2013). Applying competing risks regression models: an overview. *Lifetime Data Analysis*, 19(1):33–58.

Jeong, J.-H., and Fine, J. (2006). Direct parametric inference for the cumulative incidence function. *Journal of the Royal Statistical Society: Series C: Applied Statistics*, 55(2):187–200.

Kalbfleisch, J. D., and Prentice, R. L. (2002). *The statistical analysis of failure time data*, volume 360. John Wiley & Sons, New York.

Larson, M. G., and Dinse, G. E. (1985). A mixture model for the regression analysis of competing risks data. *Journal of the Royal Statistical Society: Series C: Applied Statistics*, 34(3):201–211.

Lau, B., Cole, S. R., Moore, R. D., and Gange, S. J. (2008). Evaluating competing adverse and beneficial outcomes using a mixture model. *Statistics in Medicine*, 27(21):4313–4327.

Makeham, W. (1874). On an application of the theory of the composition of decremental forces. *Journal of the Institute of Actuaries*, 18(5):317–322.

Mariani, L., Coradini, D., Biganzoli, E., Boracchi, P., Marubini, E., Pilotti, S., Salvadori, B., Silvestrini, R., Veronesi, U., Zucali, R., et al. (1997). Prognostic factors for metachronous contralateral breast cancer: a comparison of the linear cox regression model and its artificial neural network extension. *Breast Cancer Research and Treatment*, 44(2):167–178.

Mitchell, T. M. (1997). *Machine Learning*. McGraw-Hill series in computer science. McGraw-Hill, 1st edition.

Moeschberger, M., and David, H. (1971). Life tests under competing causes of failure and the theory of competing risks. *Biometrics*, 27:909–933.

Pintilie, M. (2006). *Competing risks: a practical perspective*, volume 58. John Wiley & Sons.

Prentice, R. L., Kalbfleisch, J. D., Peterson Jr, A. V., Flournoy, N., Farewell, V. T., and Breslow, N. E. (1978). The analysis of failure times in the presence of competing risks. *Biometrics*, 34(4):541–554.

Putter, H., Fiocco, M., and Geskus, R. B. (2007). Tutorial in biostatistics: competing risks and multistate models. *Statistics in Medicine*, 26(11):2389–2430.

Safavian, S. R., and Landgrebe, D. (1991). A survey of decision tree classifier methodology. *IEEE Transactions on Systems, Man, and Cybernetics*, 21(3):660–674.

Sreedevi, E., and Sankaran, P. (2013). Analysis of competing risks data using neural network models. *International Journal of Statistics and Applications*, 3(4):123–131.

Tsiatis, A. (1975). A nonidentifiability aspect of the problem of competing risks. *Proceedings of the National Academy of Sciences*, 72(1):20–22.

Wang, P., Li, Y., and Reddy, C. K. (2019). Machine learning for survival analysis: a survey. *ACM Computing Surveys (CSUR)*, 51(6):1–36.

Zupan, B., Demšar, J., Kattan, M. W., Beck, J. R., and Bratko, I. (2000). Machine learning for survival analysis: a case study on recurrence of prostate cancer. *Artificial Intelligence in Medicine*, 20(1):59–75.

8 PeNLP Parser

An Extraction and Visualization Tool for Precise Maternal, Neonatal and Child Healthcare Geo-locations from Unstructured Data

Patience Usoro Usip, Moses Effiong Ekpenyong, Funebi Francis Ijebu, Kommomo Jacob Usang, and Ifiok J. Udo
University of Uyo, Nigeria

CONTENTS

DOI: 10.1201/9781003161233-8

8.1   INTRODUCTION

Large disparities in maternal and neonatal mortality distinguish low-income from high-income countries (Roder-DeWan et al. 2020). One valid reason that supports this distinction is the current structure or design of health systems. Unlike high-income countries, a substantial proportion of births recorded in low-income countries occur in primary care facilities that offer no care support to complicated cases. Improving the quality of maternal and child healthcare is therefore a sure path to reaching the Sustainable Development Goals (SDGs) and universal health coverage (Goyet et al. 2019). Numerous studies have shown that not all maternal and child health indicators rely on facility-based service provision. Spatial analysis of these indicators has drawn up novel distribution of natural and cultural features, including the very "context of place" (Bassey and Akinkunmi 2013; Callaghan 2014), which extends beyond physical and political boundaries, to incorporate the social determinants of health. Indicators of health behavior in homes and communities may include processes such as exclusive breastfeeding, recommended dieting, healthcare seeking behaviors, and adherence to prescribed medical treatments; and support the need for spatial techniques deployment, for the indentation of where support is needed. Spatial analysis can also facilitate the identification of the indented areas, aiding precise allocation of resources and real-time interventions for effective healthcare provision. Distance from the nearest health facility may also pose significant barrier to seeking care for child illness or during emergency situations, for instance, due to the direct cost of travel and the indirect cost of time lost during travel (Bruce et al. 2014; Kassile et al. 2014; Bennett 2015). Harnessing geospatial technology can assist locate hotspots of low coverage and the visualization of facilities earth, to determine where community health workers are most needed and where health programs can have the most needed impact, including the prompt deployment of scarce resources. Hence, the benefits of healthcare surveillance for identification of high profile or hotspots of low coverage with high demand can be deployed for effective resource allocation, sufficient to drastically reduce health inequities between and within countries.

8.1.1   LOCATION INFORMATION MINING

Clinical notes and health records are the necessary pointers to deriving accurate locations and related temporal features of patients. These notes and records are usually represented in an unstructured format and contain spatial or location information of patients. Texts in addition to maps and satellite images are becoming important spatial data resource for representing spatiotemporal attributes of two aspects of physical appearance–space and time. Within the health sector, decision making constitutes the fulcrum for strong public health policies formulation in support of appropriate SDGs. Although the application of the geographic information system (GIS) to MNCH have flourished the literature (Molla et al. 2017; Salehi and Ahmadian 2017), only few works directly apply clinical data or patients' health records. One major factor that limits research in this area is the exclusive transforming of (semi/unstructured) data to structured form for intelligent mining can only be done by the health workers, due to confidentiality and ethical issues. The usual preprocessing steps for extracting patients' health records may include a scan of the notes or records in its textual form–to obtain the required data in a structure familiar to secondary users. Hence, a natural language parser (Mazidi and Tarau 2016) is required to mine the unstructured records and bridge the lacuna created between medical records, physicians, and researchers.

Web documents such as news articles, social feeds, and blogs provide an abundant and readily available source of spatial information related to dynamic geographic events such as wildfires, storms, and riots. Research based in the fields of geographic information retrieval (GIR) and natural language processing (NLP) use methods to extract place names and other spatial references from web documents that can be used to display event locations in a GIS. In low- and medium-income countries such as the sub-Saharan Africa, most clinical data of maternal, neonatal and child health records are largely paper based. Aside the manual storage of these records, those migrated to digital platforms are still largely unstructured. With the advent of Big Data and the increasingly large volume of unstructured but useful data available on the secure and unsecure web including social media, and the complexity associated with differing presentations of the same or similar information; it becomes highly intractable to extract useful information from even the same source. If the information about a patient or health facility cannot be extracted automatically through some computational means; then a geocode of health information (patient's location, health facility, spatial guide) is incomplete. In Adnan and Akbar (2019), they opine that supervised learning, deep learning and transfer learning techniques are the most suitable techniques. They suggest large datasets for the efficiency of these methods to be clearly observed.

Place description is a conventional recurrence in conversations involving place recommendation and person direction in the absence of a compass or a navigational map. A place description provides locational information in terms of spatial features and the spatial relations between them. When place descriptions are verbally performed, automatic extraction of spatial features is near-impossible due to non-satisfaction of locative expression requirements. But when such place descriptions are available in a natural language text; the location can easily be extracted because of the unavoidable prepositional inclusion in the written description. The inclusion of preposition before location naming and description is referred to as locative expression (Khan et al. 2013) and is easily identifiable with NLP techniques. Existing research progress in NLP have evolved human intelligent parsers with preposition-enabled natural language processing (PeNLP) capabilities (Usip et al. 2017) and is proposed in this work. To answer pertinent questions from any clinical note or record, the reports are best given in sentential form. Our interest lies in obtaining locational/

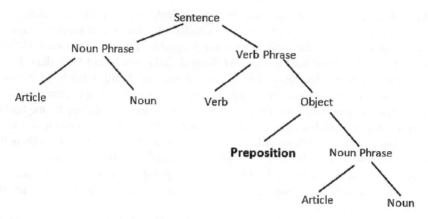

FIGURE 8.1   A sentence parse tree.

place-related terms, which are introduced by prepositions in English language, the universal language for learning. A sentence is known to have a subject, verb and object phrase as represented in the sentence parse tree in Figure 8.1, where preposition introduces the noun phrase in the object.

The proposed PeNLP parser draws its strength from the sentence parse tree. Despite the ambiguous nature of Preposition, its types can point to place terms in any given sentence, thereby enabling us to extract locational terms from any given sentence being the input to our parser.

8.1.2   OBJECTIVES AND CONTRIBUTIONS TO KNOWLEDGE

The overarching goal of this chapter is to implement a PeNLP parser for maternal, neonatal and child health information extraction from secure and unsecure websites including social media independent of the complexity of the text information, so long as it is expressed in a natural language (English). The specific objectives include:

- To extract spatial information from unstructured-clinical records of maternal, neonatal and child patients–for intelligent location information mining.
- To transform the extracted spatial data into a digitized form–for efficient mining of maternal, neonatal and child records.
- To construct a robust PeNLP tool that integrates the transformed spatial dataset– for patient-enabled query and visualization of health-specific information–for proper location-based information management and policy decision making.

The contributions of this chapter to knowledge include:

- *Interactive Open-source Tool*–Open-source tools are important remedies for improving the dynamism of health information records anywhere, anytime. Our proposed tool offers a dynamic framework that enhances an interactive tool for prompt visualization of spatial resources and knowledge simplification–through improved community contributions and prospects for end-user's feedback/recommendation.

- *Cross-border Maternal, Neonatal and Child Health (MNCH) Services Coverage–*Cross-border patterns of MNCH services coverage are uniquely identified by spatial autocorrelation analysis. This work provides excellent health-related services that address persistently low coverage of maternal and child health indicators, hence, opening further research opportunities for intelligent spatial data analysis.
- *Real-time Visualization of Health-specific Information–* The global community has coordinated multilateral regional responses to malaria, HIV, Ebola virus disease, including other outbreaks and health conditions; similar efforts This research explains the underlying reasons for low coverage areas with intense disease burden, for target-specific resource management.
- *Spatial Analysis of Comprehensive MNCH Indicators–*Visualizing spatial patterns of MNCH indicators across countries and their sub-regions is necessary to drive an efficient assessment of these indicators across national boundaries. This research provides a map-based visualization of MNCH indicators at sub-regional-level exploratory spatial data analysis (ESDA) for each indicator as well as a presentation of spatial clusters, and characterization of clusters, to account for the relationships among neighboring sub-regions. Furthermore, limitations placed by geographic barriers challenging the coverage of key MNCH through access to facility-based services are removed.
- *Adaptable System Framework–*The proposed method is adaptive to other geographical and social contexts.
- *Availability of clinical data for Sub-Saharan Africa (SSA)–*Availability of clinical data is important for the development of evidence-based health information systems. This research has built a resource-base for research continuity in the SSA. The research augments the body of established research that identifies and explores spatial implications for MNCH with a new visualization of subnational population of key MNCH indicators.

8.2 RELATED WORKS

8.2.1 Natural Language Processing of Clinical Data

The intense production of unstructured clinical content has informed the need to transform free text data into a more structured form. While unstructured data do not have a predefined schema, structured data (Maham et al. 2017) are meticulously organized and can easily be ordered and processed. NLP algorithms are helpful at locating unstructured clinical data embedded in free text notes and have been deployed successfully in pharmacogenetic studies to extract medication history from clinical narratives. The NIH-funded i2b2 initiative (Informatics for Integrating Biology and the Bedside), for instance, uses NLP software to extract clinical data from existing datasets, which are then combined with genomic data for the purpose of designing personalized medicine for patients with genetic diseases. Rangasamy et al. (2018) defines NLP as a specialized brand of AI focused on interpretation and manipulation of human-generated content be it spoken or written. It is a rapidly developing area of machine learning that helps to solve the unstructured data problem. Hence, NLP, will not only reduce access time but will assist manual expert reviews–as healthcare professionals will not only spend less time reading and interpreting electronic heath records (EHR) and free texts, but also benefit from large scale automation processing. Novel approaches that complement and go beyond evidence-based medicine are required in the domain of chronic diseases, given the growing incidence of such conditions on the worldwide population.

A promising avenue therefore is the secondary use of EHRs, where patients' data, are analyzed for clinical and translational research. Methods based on machine learning to process EHRs are improving understanding of patient clinical trajectories and chronic disease risk prediction, creating a unique opportunity to derive previously unknown clinical insights. However, a wealth of clinical histories remains locked behind clinical narratives in free-form text. NLP technologies have played a crucial role as much of detailed patient information in EHRs is embedded in narrative clinical documents. Meanwhile, several clinical NLP systems, such as: MedLEE (Sevenster et al. 2012), MetaMap/MetaMap Lite (Demner-Fushman et al. 2017), cTAKES (Savova et al. 2010), and MedTagger (Liu et al. 2013) have been developed and utilized to extract useful information from diverse types of clinical text, such as clinical notes, radiology reports, and pathology reports. Success stories in applying these tools have been reported widely. Despite the demonstrated success of NLP in the clinical domain, methodologies and tools developed for the clinical NLP are still under known and under-utilized by students and experts in the general NLP domain, mainly due to the limited exposure to EHR data.

NLP tools have been found to identify key syntactic structures in free text and extract the meaning behind the narrative. The results can then be used to generate new documents in the form of summary or be translated into codes for billing purposes. The unique content and complexity of clinical documentation can be challenging for many NLP developers, but keen interest in emerging machine learning and artificial intelligence strategies are helping to refine the industry's information processing capabilities. Healthcare natural language processing uses specialized engines capable of crowdsourcing large sets of unstructured health data to discover previously missed or improperly coded patient conditions. Natural language processing of medical records using machine learning algorithms can therefore uncover disease (s) that may not have been previously coded. NLP technologies have found applications in the automatic extraction and encoding of clinical information from narrative clinical notes. MedTagger for instance, is a knowledge driven clinical NLP system which enables sentence detection, word tokenization, section identification, contextual information, and concept identification. NLP applied to clinical narrative may overcome the limitations of billing code algorithms for identification of clinical information by recognition of text which describes signs and symptoms used to establish a diagnosis. Previous studies however demonstrate the superiority of NLP methods over billing code algorithms for phenotype identification of narrative clinical notes from the EHR.

Afzal et al. (2018) for instance extend a previously validated NLP algorithm for peripheral artery disease (PAD) identification to develop and validate a sub-phenotyping NLP algorithm (CLI-NLP) for identification of CLI cases from clinical notes. Khachidze et al. (2016) introduced the instrument for medical records classification based on the Georgian language. About 24,855 examination records were studied and classified into three main groups (ultrasonography, endoscopy, and X-ray) and 13 subgroups using two well-known methods: Support Vector Machine (SVM) and k-Nearest Neighbor (kNN). Their results demonstrate that both machine learning methods performed successfully, with SVM maintaining the best performance. Cedeño-Moreno and Vargas-Lombardo (2018) developed a software architecture using natural language processing tools and the use of an ontology of knowledge as a knowledge base. Their software extracts, manages and represents the knowledge of a text in natural language. A corpus of more than 200 medical domain documents from the general medicine and palliative care areas was validated, demonstrating relevant knowledge elements for physicians. Indicators for precision, recall and F-measure, were then applied, with validation methods yielding 95.56%, 88.56% and 91.93%, respectively, indicating an average overall accuracy of 90%. Mykowiecka et al. (2009) described a rule-based information extraction (IE) system developed for Polish medical texts. They present two applications designed to select data from medical documentation in Polish: mammography reports and hospital records of diabetic patients.

Hybridized approaches have also been explored to complement the NLP approach. Zhao (2019) for instance developed deep-learning–based NLP algorithms for automatically extracting biomarker status of patients with breast cancer from three oncology centers in Bulgaria. used dual embeddings for English and Bulgarian languages, encoding both syntactic and polarity information for the words. The embeddings were subsequently aligned so that they were in the same vector space. We showed that we can resolve ambiguity in highly variable medical text containing both Latin and Cyrillic text. Final models incorporating both English and Bulgarian syntax and polarity embeddings achieved F1 scores of 0.90 or higher for all estrogen receptor, progesterone receptor, and human epidermal growth factor receptor two biomarkers. Kang et al. (2017) developed an open-source information extraction system called Eligibility Criteria Information Extraction (EliIE) for parsing and formalizing free-text clinical research eligibility criteria (EC) following Observational Medical Outcomes Partnership Common Data Model (OMOP CDM) version 5.0. EliIE parsed EC in four steps: (1) clinical entity and attribute recognition, (2) negation detection, (3) relation extraction, and (4) concept normalization and output structuring. Informaticians and domain experts were recruited to design an annotation guideline and generate a training corpus of annotated EC for 230 Alzheimer's clinical trials, which were represented as queries against the OMOP CDM and included 8,008 entities, 3,550 attributes, and 3,529 relations. In task-specific evaluations, the best F1 score for entity recognition was 0.79, and for relation extraction was 0.89. The accuracy of negation detection was 0.94. The overall accuracy for query formalization was 0.71 in an end-to-end evaluation.

8.2.2 Spatial Knowledge Extraction from Clinical Data

One of the features of unstructured clinical data is that it contains spatial information. Texts in addition to maps and satellite images, have become an important spatial data resource in recent years (Zenasni et al. 2018). The ability to identify locations in unstructured text and quickly generate a visualization in the form of map, unlocks valuable information about the context of the locations in the text (Gehring 2015). Within the larger field of NLP, new methods are being developed to integrate geographic information with basic texts and human language. Methods that employ named entity recognition (NER) have enabled improved methods for automatically finding relevant place names. Recent developments in Artificial Intelligence (AI) and machine learning have simplified the detection of place names in unstructured texts where data can be parsed to glean important information about a place in discussion. GeoTxt, a tool which uses NER techniques and models, is often deployed to automatically finding proper names and other nouns and associating them with given places or objects as required (Karimzadeh et al. 2019).

Instances where algorithms can learn and determine the place conveyed in the information include key heuristics and *toponym* resolution methods such as supervised machine learning, which often deploy statistical and patterned behavioral rules to suggest a likelihood of a given place being in discussion (Karimzadeh et al. 2019). Other studies have also confirmed that statistical-based supervised learning approaches, like GeoTxt, enable easier auto-recognition of spatially relevant place names. However, challenges are evident, they have included developing methods that can better capture natural language change, spelling variation, and different types of language concepts conveyed in text in parsing and determining the location of places within text. In other words, by heavily depending on statistical supervised learning methods, training data that does not fully represent varied but relevant cases are likely to miss identifying relevant places. Combining multiple and disparate datasets with different training procedures could be one way such cases are resolved (Won et al. 2018).

Methods have developed gazetteers and models with NER methods that could then be deployed for an analysis. More efficient methods are generally those that combine multiple sets of NER models and gazetteers, combing the strengths of approaches they have learned from previous results (Hu et al. 2019). Other developments include disambiguating textual data referring to the same place. Sometimes, multiple texts might be discussing the same place, thereby leading to complications in disambiguating when texts are reporting the same, nearby, or more distant places due to the descriptors used. However, to match descriptions and compare them to each other allows for an ambiguous scale development. This means that comparing contextual data, and not just the place name, determines if an area discussed in one text is discussing same or different place in another text or even within the same text. These place descriptions are then compared to create and develop what is known to be spatial relations (Kim et al. 2017).

Geographic information systems are useful to establish spatial relations between health indicators and to support real-time health information management. They define geographic inequalities in service provision and guarantee informed planning decisions. Maximizing the value of spatial health access therefore requires a complete census of health providers and their locations. Albacete et al. (2012) developed a method for homebuyers, adaptable to environmental variables of interest to homebuyers when selecting a home location. They adopted a multicriteria spatial analysis method to demonstrate the homebuyers' selection process, using data from the City of Kuopio, Finland. Several spatial variables were applied, including environmental and service factors in the home searching process. A GIS was then deployed to create maps for decision variables and to map suitable areas. The method for ranking alternative dwellings was based on the difference between levels of the decision variables for each dwelling and the target levels given by the user. Roder-DeWan et al. (2020) proposed that health systems need to be redesigned to shift all deliveries to hospitals or other advanced care facilities, to bring care in line with global best practice. Health system redesign will require investing in high-quality hospitals with excellent midwifery and obstetric care, boosting quality of primary care clinics for antenatal, postnatal, and newborn care, decreasing access and financial barriers, and mobilizing populations to demand high-quality care. Maina et al. (2019) assembled national master health facility lists from a variety of government and non-government sources from 50 countries and islands in Sub-Saharan Africa and used multiple geocoding methods to provide a comprehensive spatial inventory of 98,745 public health facilities. Khachidze et al. (2016) automatically extracted species names of bacteria and their locations from webpages. They designed a new model for joint extraction of biomedical entities and the localization relationship. Their model was based on a spatial role labeling (SpRL) model designed for spatial understanding of unrestricted text. Furthermore, they exploited the vast amount of biological knowledge expressed in diverse natural language texts and put this knowledge in databases for easy access by biologists.

8.2.3 THE LINGUISTIC AND LOGICAL THEORY OF PLACE

Place is a fundamental concept in geography and plays a key role in almost every field of human inquiry (Tuan 1990; Harrison and Dourish, 1996, Jordan et al. 1998). 'Place', despite being the basic notion in everyday communication, appear controversial in its semantic understanding. However, the linguistic and logical theory of place (Agarwal 2005), proposed four important categories of linguistic expression relating to place as follows: Count Nouns, Locative Property Phrases, Place Names and Definite Descriptions. The prepositions adopted by PeNLP parser points to this categorization for place identification during unstructured text parsing. The challenge of accurately defining the space of a given place can therefore be resolved within the domain of discuss. Given a city within a defined neighborhood, several events relevant to the city are bound to occur. The use of text and comparing different gazetteers do not provide enough evidence in defining the spatial dimension. Hence, the dimensions of a given discuss may be extracted or learned and require methods for searching word descriptions that matches approximate measures of area. As such, investigations and extractions of contexts that define descriptive terms in relation to wider space are most relevant for the place in discussion (Paterson and Gregory Ian, 2019).

8.3 MATERIALS AND METHODS

The following subsections discuss the data source, system architecture, parser design and extraction methods deployed to achieve our study objectives.

8.3.1 DATA SOURCE

The study area of this research is shown in Figure 8.2. After ethical approval from the University of Uyo Institutional Health Research Ethics Committee (UNIUYO-IHREC) and the Saint Luke's Hospital Management, clinical data of patients were collected with the assistance of health workers, from the patient's health records for maternal, neonatal and child health units of St Luke's General Hospital, Anua, Uyo. For this paper, patient's data related to their location and disease condition/ailment were excavated over a period of six years (2014–2020). The semi-structured data presented here were drawn from the unstructured notes by the clinical staff (due to confidentiality issues) under the watch and guidance of the research assistant in the project team. A total of 40 unique sample points each were processed for maternal, neonatal and child health data as represented in Tables 8.1–8.3, respectively. The patients-IDs are given as Pm1...Pm30 for maternal cases, Pn1...Pn30 for neonatal cases, and Pc1...Pc30 for child cases.

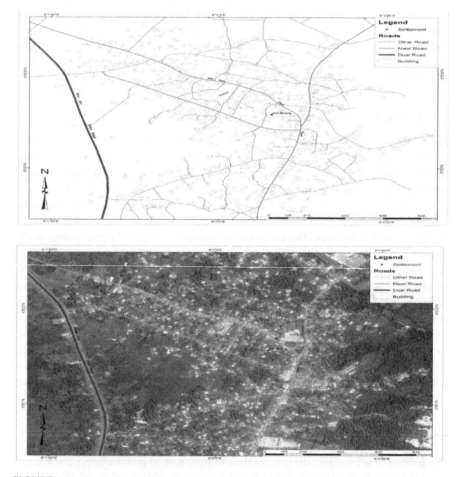

FIGURE 8.2   Study area showing deployment environment.

TABLE 8.1
Maternal Locational Data with Ailment from 2014 to 2020

S/NO.	Patient's ID	Location	Ailment
1	Pm1	Ikot Essien Ibiono	URTI
2	Pm2	Ekritam Akpan Obong, Itu	Pre-eclampsia
3	Pm3	Nkim Itam, Itu	Malaria
4	Pm4	Udi Street, Off Abak Road	Glucosuria
5	Pm5	Ekpri Nsukara	Malaria
6	Pm6	Ikot Ambang Ibiono	Proteinurea
7	Pm7	Itiam Street	Raised BP
8	Pm8	Akamba Nsukara	URTI
9	Pm9	Ikpa Road	PPA
10	Pm10	Nwaniba Road	Oedema
11	Pm11	Brook Street	Asthma
12	Pm12	Itu Road	HTN
13	Pm13	Inyang Street, Itam	Diarrheoa
14	Pm14	Ikot Andem Itam	Retained Placenta
15	Pm15	Okon Essuene	Hepatitis
16	Pm16	Urua Ekpa	UTI
17	Pm17	Idu Uruan	HTN
18	Pm18	Abak Road	Typhoid Malaria
19	Pm19	Ikpa Road	Typhoid Malaria
20	Pm20	Obot Idim Ibesikpo	Diarrheoa
21	Pm21	Abak Road	Normal Booking
22	Pm22	Oron Road	Retained Placenta
23	Pm23	Nelson Mandela	Chicken Pox
24	Pm24	IBB Avenue, Uyo	GLYCOSUREA
25	Pm25	Idak Okpo, Uyo	Ulcer
26	Pm26	Barracks Road, Uyo	Normal Booking
27	Pm27	Idu Uruan	Chicken Pox
28	Pm28	Obio ImoStreet, Uyo	Malaria
29	Pm29	Ikot Ekpene Road, Uyo	Normal Booking
30	Pm30	Ebong Umoitong Street, Off Atiku A. Way	Hyperemesis
31	Pm31	Obot Street, Itam, Itu	Liquor Drainage
32	Pm32	AkpaIkpe Street, Uyo	Haematuria
33	Pm33	Out Street, Nung Oku	Eclampsia
34	Pm34	Osongama Housing	Malaria
35	Pm35	Atiku Abubakar, Uyo	Normal Fetal Scan
36	Pm36	Esuene street	Normal Booking
37	Pm37	Eda Street, Anua	Diarrheoa
38	Pm38	UdoUmana Street	Retained Placenta
39	Pm39	Aka Road, Uyo	Malaria
40	Pm40	Mbierebe Obio	Normal Booking

Source: Health Records, St. Luke's General Hospital, Anua.

TABLE 8.2

Neonatal Locational Data with Ailment from 2014 to 2020

S/NO.	Patient's ID	Location	Ailment
1	Pn1	Udo Eduok Street	Premature
2	Pn2	Attan Offot Road	NNJ
3	Pn3	Afaha Ube Street	Sepsis
4	Pn4	Oron Road	NNJ
5	Pn5	Ekit Itam	NNS
6	Pn6	Bassey Utuk	Preterm
7	Pn7	Essien Essien Street	SBA
8	Pn8	Oron Rd	NNS
9	Pn9	Ikot Ekwere Road, Itam	Congenital Pnuemonia
10	Pn10	AKPKPAN Street	MBA
11	Pn11	Mbiokporo I, Nsit Ibom	NNS
12	Pn12	1000 Unit, Idu Uruan	NNJ
13	Pn13	Ifa Ikot Okpon	NNS, Congenital Phuemonia
14	Pn14	Abak Road	NNJ, Resolving Sepsis
15	Pn15	Udo Idiong Nka Lane	Fetal Macrosomic
16	Pn16	Osongama Road	NNJ Resolving
17	Pn17	Ayadehe Itu	Macrosomic Neonate
18	Pn18	Urua Ekpa Road	PLBW
19	Pn19	Aka Etinan Road	Macrosomic
20	Pn20	Idoro Road	NNJ
21	Pn21	High Tension	NNS
22	Pn22	4- Lane, Ewet Housing	LBW
23	Pn23	Etim Okon Usanga	SBA
24	Pn24	Ikot Ekpene Road	Post-term Infant with MAS
25	Pn25	Aka Road	Constitutional Macrosomia
26	Pn26	Sunday Ibanga Street	Presumed Sepsis
27	Pn27	Ibiaku Issiet, Uruan	Severe Birth Asphyxia
28	Pn28	Effiong Inim Street, Ifa Atai	Meconium Aspiration
29	Pn29	Udoette Street	Liquor Aspiration
30	Pn30	Effiong Inim Street	Moderate Birth Asphyxia
31	Pn31	Aka Idak Eyop	Congenital pneumonia
32	Pn32	Mbribit Itam	Congenital Pnuemonia
33	Pn33	Efiat street	Congenital Pnuemonia
34	Pn34	Nung Oku Ibesikpo	Exposed Baby
35	Pn35	Aka Etinan Road	Exposed Baby
36	Pn36	Akpan Essien Street	Low Apgar Score
37	Pn37	Utuks Lane	Congenital Missing Tibia
38	Pn38	Ikot Akpan Abia	Retroviral Disease Syndrome
39	Pn39	Mbak Itam	Observation
40	Pn40	Mbiakpan Atan	Opthalmia Neonate

Source: Health Records, St. Luke's General Hospital, Anua.

TABLE 8.3
Child Locational Data with Ailment from 2014 to 2020

S/NO.	Patient's ID	Location	Ailment
1	Pc1	Ndon Ebom, Uruan	Febrile convulsion
2	Pc2	Asikpo Street, Ekpri Nsukara	Severe Malaria
3	Pc3	Etim Okon Usanga	Acute Plasmodiasis
4.	Pc4	Mbikpong Ikot Edim, Ibesikpo	Abdominal Distension
5	Pc5	Bassey Akpan Street, Mbiabong	Abdominal Colic
6	Pc6	Nung Ikono, Uruan	Acute Watery Diarrhea
7	Pc7	Akamba Nsukara Offot	Scistica
8	Pc8	Unit B Ewet Housing Estate	Otitis Media
9	Pc9	Nwaniba Road	Acute plasmodiasis
10	Pc10	Mbiabong Ikot Enniang, Ini	Umbilical Hernia
11	Pc11	Ritman Rd., Ikot Ekpene	RTA
12	Pc12	Osongama, Off Abel Damina Str	Otitis Media
13	Pc13	Archibong Street Mbiabong Etoi	URTI
14	Pc14	Ukana Ebet Lane, Off IBB	RTI
15	Pc15	Peter Uboh, Nkemba Street	Bronchopnuemonia
16	Pc16	Use Offot	Severe Malaria
17	Pc17	Nepa Line, Nsukara	Scabies
18	Pc18	Ifa Atai	Chest Pain Trauma
19	Pc19	Out Udosen Street, Ekpri Nsukara	Acute Tosilitis
20	Pc20	Ikot Obio Ndua, Nsit Ubium	Hydrocele
21	Pc21	Ifa Atai, Uyo	Rhinorrhea
22	Pc22	Atan Ikot Abia Edim	Allergic Rhinitis
23	Pc23	Afaha Nsit	Severe disorder and cerebral palsy
24	Pc24	Utang Street, Uyo	Sepsis
25	Pc25	100 Unit, Udoudoma	Severe Anaemia
26	Pc26	School Road, Itu Road	Dysentery
27	Pc27	Shelter Afrique	Severe Anaemia
28	Pc28	Ring Road, Mbiabong	Acute Watery Diarrhea
29	Pc29	Thomas Udo Ekong, Off New Stadium	Febrile Convulsion
30	Pc30	Ravine Estate, Eniong Offot	Trauma to the ear
31	Pc31	Anua Street, Abak Road	ARI
32	Pc32	Ifiayong Usuk, Uruan	Acute Kwoshokor
33	Pc33	Oron Road	Gastroentiritis
34	Pc34	Mbiakpan Atan, Ibiono	Malaria
35	Pc35	Ikot Ebinyang, Etinan	Frequent Stooling
36	Pc36	Itak junction	Plasmodiasis
37	Pc37	Abak Road	GET
38	Pc38	Silas Street, Use Offot	BPN
39	Pc39	Eman Road, Idu Market	Indigestion
40	Pc40	Akpasak Estate	NNJ

Source: Health Records, St. Luke's General Hospital, Anua.

8.3.2   THE PeNLP PARSER ARCHITECTURE

The current PeNLP parser architecture (Figure 8.3) modifies Usip et al. (2017), and represents the processes in the PeNLP parser algorithm described in Figure 8.4. Health records/clinical notes in the health facilities provide the required answers to the competency questions as basis for the unstructured text used as input to the parser, which approach is discussed in the following subsections.

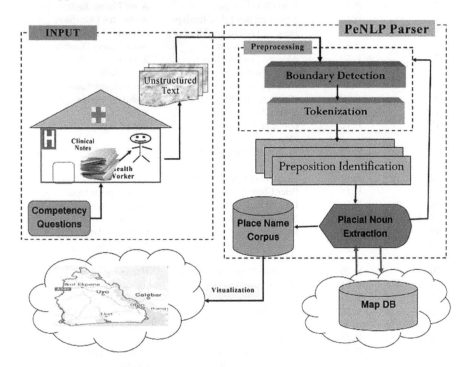

FIGURE 8.3   The PeNLP parser architecture.

```
get unstructured text
split unstructured text into Sentences using full stop as boundary marker
locate finite Verb in Sentence
        if no finite Verb then
                locate Auxiliary Verb
                split Sentence into Subject and Predicate
                if Preposition exists in Subject then
                        get Word or Phrase
                until "," or Auxiliary Verb or "?,!"
                else
                if there is Preposition in Predicate, then
                        get Word or Phrase
                until "," or Auxiliary Verb or "?,!"
                assign Phrase or Word to Spatial Context or Placial Noun
                store Placial Noun or Spatial Context
                match Placial Noun with precise Geo-location in the Map database
                visualize the location
repeat until Sentence = "|"
```

FIGURE 8.4   The modified PeNLP parser algorithm.

8.3.3 THE PENLP PARSER

Preprocessing techniques are applied to the raw data using a PeNLP tool as proposed in Maham et al. (2017), where prepositions in grammar are used to identify the spatial concepts in any given unstructured text. The proposed approach splits statements into sentences using a boundary marker (.) and the existence of verbs (main verb) in each sentence is searched for and in the absence of a main verb, the auxiliary verb is retrieved before breaking the sentence into a subject and a predicate. The algorithm checks for prepositions in both the subject and predicate. On finding the prepositions, the word or phrase is extracted as the placial nouns. This process is repeated for all sentences and the placial nouns stored as spatial knowledge in the repository. The representation and classification of placial nouns/spatial context is based on Agarwal's structure.

The PeNLP parser approach consists of several technical processes including:

8.3.3.1 Sentence Construction

For uniformity, the collected unstructured data set are answers to the following question-set with the first question in the question-set giving rise to the sentences considered as input to the PeNLP parser: The question-set includes:

- Where is the PATIENT's *residence*?
- What was the *diagnosed disease condition* of the PATIENT, *cause* and *administered treatment*?
- *Where* does the PATIENT spend most his or her time (e.g., *office, business premises*, etc.)?
- Which *health facility* did the PATIENT visit and where is it *located*?
- What *date* did the PATIENT visit the *health facility*?
- Are there other *health facilities* close to the PATIENT's *residence*?
- What were the *reported symptoms* on the *date* of visit by PATIENT?
- What was the *diagnosed disease condition* of the PATIENT, *cause* and *administered treatment*?

8.3.3.2 Sentence Boundary Detection

Sentence boundary detection is based on the following regular expression that defines a sentence with <Noun> as an additional component to the sentence expression defined based on Agarwal's classification of place terms and given as:

<Sentence> = <Subject> <Predicate>

<Subject> = <Noun Phrase>

<Predicate> = <Verb> <Object>

<Object> = < Preposition> <Noun Phrase>

<Noun Phrase> = <Article> <Noun>

<Noun> = <Geolocation> <Location Description> <Generic Place Names>

8.3.3.3 Tokenization

Tokenization involves locating the component parts of the parsed sentences. First, the subject and predicate are located using the verb as the keyword. Also, all white spaces and stop-words are removed. As a follow up, prepositions are identified, followed by the noun phrase as placial phrase, where one placial phrase may contain more than one placial noun. Hence, given the following sentence:

"Patient *Pn38* diagnosed with Retroviral Disease Syndrome, resides at Ikot Akpan Abia."

The resultant tokens are broken into:

Subject: *'Patient Pn38 diagnosed with Retroviral Disease Syndrome'.*

Predicate*: 'resides at Ikot Akpan Abia'.*

Verb: *resides*

Object: *at Ikot Akpan Abia.*

Preposition: *at.*

Noun Phrase: *Ikot Akpan Abia.*

Noun: *Ikot Akpan Abia.*

8.3.3.4 Lexical Matching/Preposition Identification

This process considers the set of English keywords known as prepositions, enumerated in the PeNLP parser preposition-set in Figure 8.5. The search for prepositions in the bounded parts of any sentence is done by matching the keyword before the <*Noun Phrase*> in the <*Object* > by each of the keywords in the <*Preposition-Set*>. Where there is a match, then the <*Noun Phrase* > is a possible place term for extraction in the next process.

8.3.3.5 Place Term Extraction (Corpus Generation)

The possible place terms were extracted from the natural language text and sent to the database of all place terms. The resulting database, otherwise called the locational corpus, may contain terms other than the classified location or place term. This is taken back to Process (d), where further matches and identification are done to exhaust all the locations within the compound sentence. Each place term or placial noun or locational terms are then added to the corpus.

Preposition-set =
{"about ", " above ", " across ", " after ", " against ", " along ", " around ",
" at ", " because of ", " before ", " behind ", " below ", " beneath ", "beside ",
"besides ", "between ", " beyond ", " but ", " by ", "concerning ", " despite ",
" down ", " during ", " except ", " excepting ", " for ", " from ", " in ",
" in front of ", " inside ", " in spite of ", " instead of ", " into ", " like ",
" near ", " of ", " off ", " on ", " onto ", " out ", " outside ", " over ", " past ",
" regarding ", " since ", " through ", " throughout ", " to ", " toward ",
" under ", " underneath ", " until ", " up ", " upon ", " up to ", " with ",
" within ", " without ", " with regard to ", " with respect to "}

FIGURE 8.5 PeNLP parser preposition-set.

8.3.3.6 Geolocation Identification/Visualization (Linkage to Google Maps)

From the resulting location or place terms, hyperlinks are added to all geolocations for easy linkage and visualization in Google Maps. A Google Maps API with a JX browser (an API that integrates a Chromium-based browser with your Java app to process and display HTML5, CSS3, JavaScript, etc.) running in Java and using the provided key from Google was used to implement this process.

8.4 RESULTS AND DISCUSSION

In Figure 8.6, a snapshot of the placial noun extractor is presented. The user may write the unstructured text in the given text area or browse to upload a plain text file containing the unstructured data. The sample constructed sentences used to demonstrate the feasibility of our design is given in Figure 8.7. On clicking the "Parse Text button", the parser results are as presented in Figures 8.8–8.10 for neonatal, maternal and child health related data, respectively.

8.4.1 INPUT

Sample sentences were constructed and stored as a text file and used as input to the PeNLP parser. However, the input file size should not exceed 25 MB. Sample constructed sentences from the neonatal cases are given in Figure 8.7.

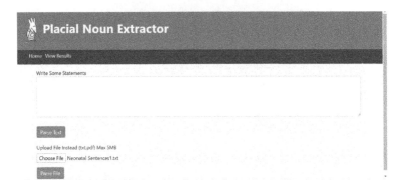

FIGURE 8.6 The placial noun (Geo-location) extractor.

> *Patient Pn11 lives at Mbiokporo I in Nsit Ibom.*
>
> *Patient Pn19 resides at Aka Etinan road.*
>
> *Patient Pn22 lives at Ewet Housing.*
>
> *At Etim Okon Usanga Street in Uyo, patient, Pn23 resides.*
>
> *Patient Pn24 is a Post-term Infant and resides at Ikot Ekpene Rd in Uyo.*
>
> *Patient Pn29 resides at Udoette street.*
>
> *Patient Pn38 resides at Ikot Akpan Abia.*

FIGURE 8.7 Sample sentences as input to PeNLP parser.

8.4.2 PENLP PARSER OUTPUTS

The generated corpus in Figure 8.8 was obtained after parsing the sample sentences in Figure 8.7. Different sets of sentences were submitted to the parser for the maternal and child health cases and the results from the parser are as shown in Figure 8.9 and 8.10, respectively.

A map generated to visualize the location place name "Udoette Street" (marked in red) using the Google Map API is shown in Figure 8.11. Google Map has proven to be the most expansive data machine and has curated extremely large linguistic resources over the years, as billions of bytes of data to map the entire world have been collected. Hence, increasing penetration of GPS-enabled smartphone has rapidly expanded Google database, to putting all these data to efficient use. Google has over the last two decades dedicated enormous resources to build a searchable map for the entire earth. They have collected more than 20 Petabytes (20 million GB) of data and are still collecting more information to constantly improve the maps. With over 1Billion monthly active users, Google Maps appears one of the most used consumer applications.

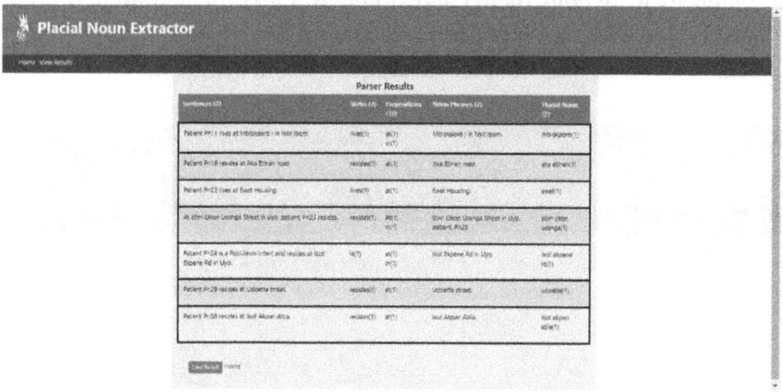

FIGURE 8.8 The placial noun corpus from unstructured neonatal data.

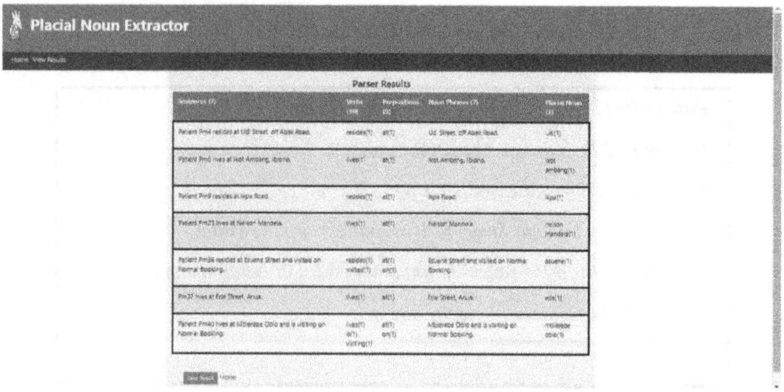

FIGURE 8.9 The placial noun corpus from unstructured maternal data.

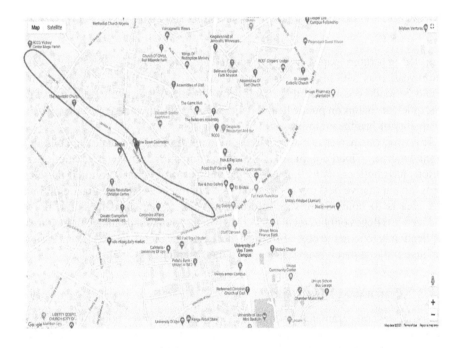

FIGURE 8.10   The placial noun corpus from unstructured child data.

FIGURE 8.11   The visualization of place name (Udoette Street) on google maps.

Figure 8.11 reveals one major limitation of Google Maps, as this huge linguistic resource is prone to errors. Occasionally, ambiguities and flaws in location data may produce a route that doesn't lead to the actual destination. Furthermore, some remote locations may not be in Google Maps, hence requiring localized mapping of the intended study area and enriching the various locations with landmark labels for proper identification. An enriched localized mapping of the current study area is

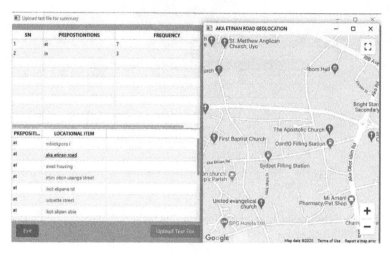

FIGURE 8.12   Placial noun and its visualization from the desktop application of the parser.

expected in a future paper, as this would increase the visibility of remote locations and reveal areas with high disease burden. Access to the respective health facilities would be made possible as these facilities will be more visible to end-users. With this, quick decisions on public health policies could be concluded more promptly and the quality of healthcare improved.

However, our parser is a web-based one with the resulting data available in the cloud database. Cloud computing is prone to competing, privacy regulations across various jurisdictions, as well as evolving cybersecurity threats. To quell these concerns, an alternative desktop application (see Figure 8.12) has been made available. The idea is to domicile private information such as patient data as clients with assigned privileges only to regulators of such data. Another advantage of the desktop application is to save the cost of frequent access to the cloud database, hence, saving the huge traffic a purely web application would manifest.

8.4.3   Performance Evaluation

Although, several parameters such as verb, prepositions, place nouns were extracted by the parser, the evaluation will be based on the place terms or locational items, as this is the only linked functionality of the map. More functionalities would be added to the proposed PeNLP and this research advances.

The performance metrics used to evaluate the PeNLP include precision, recall, and F-measure (see Equations 8.1–8.3). Computations of these metrics with respect to maternal, neonatal and child data are shown in Tables 8.4–8.6, respectively.

$$\text{Precision}, P = l_2 \,/\, l_1 \tag{8.1}$$

$$\text{Recall}, R = l_2 \,/\, l \tag{8.2}$$

TABLE 8.4

Performance Evaluation of our PeNLP for Unstructured Maternal Data

Sentence	Actual Locations in sentence, l	Locations identified by PeNLP Parser, l_1	Locations correctly linked to the map, l_2	Precision	Recall	F-measure
1	2	1	1	1	0.5	0.6667
2	2	1	1	1	0.5	0.6667
3	1	1	1	1	1	1
4	1	1	1	1	1	1
5	1	1	1	1	1	1
6	2	1	1	1	0.5	0.6667
7	1	1	1	1	1	1
Average				1	0.7857	0.8572

TABLE 8.5

Performance Evaluation of Our PeNLP for Unstructured Neonatal Data

Sentence	Actual Locations in sentence, l	Locations identified by PeNLP Parser, l_1	Locations correctly linked to the map, l_2	Precision	Recall	F-measure
1	2	1	0	0	0	0
2	1	1	1	1	1	1
3	1	1	1	1	1	1
4	2	1	1	1	0.5	0.6667
5	2	1	1	1	0.5	0.6667
6	1	1	1	1	1	1
7	1	1	1	1	1	1
Average				0.8571	0.7143	0.7619

TABLE 8.6

Performance Evaluation of Our PeNLP for Unstructured Child Data

Sentences	Actual Locations in sentence, l	Locations identified by PeNLP Parser, l_1	Locations correctly linked to the map, l_2	Precision	Recall	F-measure
1	1	1	1	1	1	1
2	2	1	1	1	0.5	0.6667
3	2	1	1	1	0.5	0.6667
4	2	1	1	1	0.5	0.6667
5	1	1	1	1	1	1
6	1	1	1	1	1	1
7	1	1	1	1	1	1
Average				1	0.7857	0.8572

$$F-\text{measure}, F_m = 2*(P*R/P+R) \tag{8.3}$$

where

   l = *actual number of locations in sentence*

   l_1 = *number of correctly identified locations*

   l_2 = *number of correctly linked locations*

F-measure is best at 1 and worst at 0. Seven sentences were considered to evaluate the proposed parser.

The best performance for maternal data was obtained for sentences 3, 4, 5, and 7 with F-measure of 100% (see Table 8.4). The worse performance however came from sentences 1, 2, and 6, with a F-measures of 66.67%. The average results gave 100% precision, 78.57% recall and 85.72% F-measure.

The best performance for neonatal data was obtained for sentences 2, 3, 6, and 7 with F-measures of 66.67% (see Table 8.5). The worse performance however came from sentences 4 and 5, with a F-measure of 66.67%. The average results gave 85.71% precision, 71.43% recall and 76.19% F-measure.

The best performance for child data was obtained for sentences 1, 5, 6 and 7 with F-measures of 100% (see Table 8.6). The worse performance however came from sentences 13, 3 and 4, with a F-measure of 66.67%. The average results gave 100% precision, 78.57% recall and 85.72% F-measure.

From the resulting place names, about 95% percent of results were successfully visualized on the map. Among the visualized place terms about 2% of the locations were spotted at countries and regions other than the study area. For instance, similar names in other regions overlapped with *Mbiokporo Nsit I* (see Figure 8.9).

In Karimzadeh, et al. (2019), the authors presented a similar work that offers six named entity recognition techniques for place name recognition on a corpus of manually geo-annotated tweets. They observed that named entity recognition algorithms trained on news stories underperformed when tested with plain text from tweets. Furthermore, even with minimal data, our parser compares favorably with the literature and can be improved upon in future paper.

8.5   CONCLUSION AND FUTURE RESEARCH PERSPECTIVE

This chapter utilized unstructured clinical records of MNCH for location extraction and visualization. The proposed tool identified and extracted locations using a preposition-enabled technique, while location visualization was achieved on Google map. Our PeNLP parser successfully parsed unstructured text to derive place terms that can be linked to Google Maps. Obtained visualizations would help in real time location of health facilities and analysis of MNCH indicators. The performance of the proposed parser was also evaluated and found to compare favorably with the literature. A future research direction is to mine Big MNCH data to perform density mapping of areas, for efficient visualization and identification of areas with high disease concentration.

ACKNOWLEDGEMENTS

This research is funded by the Tertiary Education Trust Fund (TETFund), Nigeria. Also acknowledge are the clinical staff at St. Luke's General Hospital Hospital, Anua, Uyo, and our Doctoral students for assistance they rendered during the data collection phase of this research.

DISCLOSURE STATEMENT

There is no conflict of interest.

DATA AVAILABILITY STATEMENT

The MNCH data providing the domain knowledge were obtained from Saint Luke's General Hospital, Anua, Uyo, Akwa Ibom State, Nigeria.

DECLARATION OF INTEREST

None declared.

REFERENCES

Adnan, Kiran, and Rehan Akbar. "An analytical study of information extraction from unstructured and multidimensional big data." *Journal of Big Data* 6, no. 1 (2019): 91.

Afzal, Naveed, Vishnu Priya Mallipeddi, Sunghwan Sohn, Hongfang Liu, Rajeev Chaudhry, Christopher G. Scott, Iftikhar J. Kullo, and Adelaide M. Arruda-Olson. "Natural language processing of clinical notes for identification of critical limb ischemia." *International Journal of Medical Informatics* 111 (2018): 83–89.

Agarwal, Pragya. *"Operationalising 'sense of place' as a cognitive operator for semantics in place-based ontologies."* In *International Conference on Spatial Information Theory*, pp. 96–114. Springer, Berlin, Heidelberg, 2005.

Albacete, Xavier, Kari Pasanen, and Mikko Kolehmainen. "A GIS-based method for the selection of the location of residence." *Geo-spatial Information Science* 15, no. 1 (2012): 61–66.

Bassey, Patience C., and B. O. Akinkunmi. "Introducing the spatial qualification problem and its qualitative model." *African Journal of Computing and ICTs* 6, no. 1 (2013): 191–196.

Bennett, Adam. *Global Trends in Care Seeking and Access to Diagnosis and Treatment of Childhood Illnesses*. ICF International, 2015.

Bruce, Nigel, Daniel Pope, Byron Arana, Christopher Shiels, Carolina Romero, Robert Klein, and Debbi Stanistreet. "Determinants of care seeking for children with pneumonia and diarrhea in Guatemala: Implications for intervention strategies." *American Journal of Public Health* 104, no. 4 (2014): 647–657.

Callaghan, William M. "Geographic variation of reproductive health indicators and outcomes in the United States: Place matters." *American Journal of Obstetrics and Gynecology* 211, no. 3 (2014): 278–284.

Cedeño-Moreno, Denis, and Miguel Vargas-Lombardo. "Design and construction of a nlp based knowledge extraction methodology in the medical domain applied to clinical information." *Healthcare Informatics Research* 24, no. 4 (2018): 376–380.

Demner-Fushman, Dina, Willie J. Rogers, and Alan R. Aronson. "MetaMap Lite: An evaluation of a new Java implementation of MetaMap." *Journal of the American Medical Informatics Association* 24, no. 4 (2017): 841–844.

Gehring, Sarah Marie. "Semi-Automated Visualization of Spatial Information in Unstructured Text." PhD diss., University of Southern California, 2015.

Goyet, Sophie, Valerie Broch-Alvarez, and Cornelia Becker. "Quality improvement in maternal and newborn healthcare: lessons from programmes supported by the German development organisation in Africa and Asia." *BMJ Global Health* 4, no. 5 (2019): e001562.

Harrison, Steve, and Paul Dourish. *"Re-place-ing space: the roles of place and space in collaborative systems."* In *Proceedings of the 1996 ACM conference on Computer supported cooperative work*, pp. 67–76. 1996.

Hu, Yingjie, Huina Mao, and Grant McKenzie. "A natural language processing and geospatial clustering framework for harvesting local place names from geotagged housing advertisements." *International Journal of Geographical Information Science* 33, no. 4 (2019): 714–738.

Jordan, Hans-Joachim, M. Wegner, and Hans Tiziani. "Highly accurate non-contact characterization of engineering surfaces using confocal microscopy." *Measurement Science and Technology* 9, no. 7 (1998): 1142.

Kang, Tian, Shaodian Zhang, Youlan Tang, Gregory W. Hruby, Alexander Rusanov, Noémie Elhadad, and Chunhua Weng. "EliIE: An open-source information extraction system for clinical trial eligibility criteria." *Journal of the American Medical Informatics Association* 24, no. 6 (2017): 1062–1071.

Karimzadeh, Morteza, Scott Pezanowski, Alan M. MacEachren, and Jan O. Wallgrün. "GeoTxt: A scalable geoparsing system for unstructured text geolocation." *Transactions in GIS* 23, no. 1 (2019): 118–136.

Kassile, Telemu, Razack Lokina, Phares Mujinja, and Bruno P. Mmbando. "Determinants of delay in care seeking among children under five with fever in Dodoma region, central Tanzania: a cross-sectional study." *Malaria journal* 13, no. 1 (2014): 348.

Khachidze, Manana, Magda Tsintsadze, and Maia Archuadze. "Natural language processing based instrument for classification of free text medical records." *BioMed Research International* 2016 (2016).

Khan, Hafiz, Azhar Ali, Sarfraz Ali Shad, and Waseem Akram. "Resistance to new chemical insecticides in the house fly, *Musca domestica* L., from dairies in Punjab, Pakistan." *Parasitology research* 112, no. 5 (2013): 2049–2054.

Kim, Junchul, Maria Vasardani, and Stephan Winter. "Similarity matching for integrating spatial information extracted from place descriptions." *International Journal of Geographical Information Science* 31, no. 1 (2017): 56–80.

Liu, Hongfang, Suzette J. Bielinski, Sunghwan Sohn, Sean Murphy, Kavishwar B. Wagholikar, Siddhartha R. Jonnalagadda, K. E. Ravikumar, Stephen T. Wu, Iftikhar J. Kullo, and Christopher G. Chute. "An information extraction framework for cohort identification using electronic health records." *AMIA Summits on Translational Science Proceedings* 2013 (2013): 149.

Maham, Mehdi, Mahmoud Nasrollahzadeh, S. Mohammad Sajadi, and Mehdi Nekoei. "Biosynthesis of Ag/reduced graphene oxide/Fe_3O_4 using Lotus garcinii leaf extract and its application as a recyclable nanocatalyst for the reduction of 4-nitrophenol and organic dyes." *Journal of Colloid and Interface Science* 497 (2017): 33–42.

Maina, Joseph, Paul O. Ouma, Peter M. Macharia, Victor A. Alegana, Benard Mitto, Ibrahima Socé Fall, Abdisalan M. Noor, Robert W. Snow, and Emelda A. Okiro. "A spatial database of health facilities managed by the public health sector in sub Saharan Africa." *Scientific Data* 6, no. 1 (2019): 1–8.

Mazidi, Karen, and Paul Tarau. *"Infusing nlu into automatic question generation."* In *Proceedings of the 9th International Natural Language Generation conference*, pp. 51–60. 2016.

Molla, Yordanos B., Barbara Rawlins, Prestige Tatenda Makanga, Marc Cunningham, Juan Eugenio Hernández Ávila, Corrine Warren Ruktanonchai, Kavita Singh et al. "Geographic information system for improving maternal and newborn health: Recommendations for policy and programs." *BMC Pregnancy and Childbirth* 17, no. 1 (2017): 1–7.

Mykowiecka, Agnieszka, Małgorzata Marciniak, and Anna Kupść. "Rule-based information extraction from patients' clinical data." *Journal of Biomedical Informatics* 42, no. 5 (2009): 923–936.

Paterson, Laura L., and Gregory Ian N. "Defining and measuring poverty." In *Representations of Poverty and Place*, pp. 1–18. Palgrave Macmillan, Cham, 2019.

Rangasamy, Suresh, Nadenichek Rosanne, Rayasam Mahi and Sozdatelev Alex. "Natural language processing in healthcare" *MicKinsey & Company healthcare systems and services*, December 6, 2018, https://www.mckinsey.com/industries/healthcare-systems-and-services/our-insights/natural-language-processing-in-healthcare. Accessed: December 6, 2018.

Roder-DeWan, Sanam, Kojo Nimako, Nana AY Twum-Danso, Archana Amatya, Ana Langer, and Margaret Kruk. "Health system redesign for maternal and newborn survival: Rethinking care models to close the global equity gap." *BMJ Global Health* 5, no. 10 (2020): e002539.

Salehi, Fatemeh, and Leila Ahmadian. "The application of geographic information systems (GIS) in identifying the priority areas for maternal care and services." *BMC Health Services Research* 17, no. 1 (2017): 1–8.

Savova, Guergana K., James J. Masanz, Philip V. Ogren, Jiaping Zheng, Sunghwan Sohn, Karin C. Kipper-Schuler, and Christopher G. Chute. "Mayo clinical Text Analysis and Knowledge Extraction System (cTAKES): architecture, component evaluation and applications." *Journal of the American Medical Informatics Association* 17, no. 5 (2010): 507–513.

Sevenster, Merlijn, Rob Van Ommering, and Yuechen Qian. "Automatically correlating clinical findings and body locations in radiology reports using MedLEE." *Journal of Digital Imaging* 25, no. 2 (2012): 240–249.

Tuan, Yi-Fu. *Topophilia: A Study of Environmental Perceptions, Attitudes, and Values.* Columbia University Press, New York1990.

Usip, Patience U., Moses E. Ekpenyong, and James Nwachukwu. *"A secured preposition-enabled natural language parser for extracting spatial context from unstructured data."* In *International Conference on e-Infrastructure and e-Services for Developing Countries*, pp. 163–168. Springer, Cham, 2017.

Won, Miguel, Patricia Murrieta-Flores, and Bruno Martins. "Ensemble named entity recognition (ner): Evaluating ner Tools in the identification of Place names in historical corpora." *Frontiers in Digital Humanities* 5 (2018): 2.

Zenasni, Sarah, Eric Kergosien, Mathieu Roche, and Maguelonne Teisseire. "Spatial information extraction from short messages." *Expert Systems with Applications* 95 (2018): 351–367.

Zhao, Boyang. "Clinical data extraction and normalization of cyrillic electronic health records via deep-learning natural language processing." *JCO Clinical Cancer Informatics* 3 (2019): 1–9.

9 Recent Trends in Deep Learning, Challenges and Opportunities

S. Kannadhasan
Cheran College of Engineering, India

R. Nagarajan
Gnanamani College of Technology, India

M. Shanmuganantham
Tamilnadu Government Polytechnic College, India

CONTENTS

9.1 INTRODUCTION

This paper discusses previous deep learning science, its implementations and recent advances in the processing of natural languages. Several domains, such as pictures, smells, text, and motion, have been successfully implemented. The methods built through research into deep learning have also had an influence on the process of natural language research [1–5]. The method has the value of systemic thought regarding posing and has a clear but effective formulation that capitalizes on recent developments in Deep Learning. A phenomenal set of developments in machine learning, and in particular deep learning techniques focused on artificial neural networks, have been seen over the past decade to enhance our capacity to create more detailed systems through a large variety of fields, including computer vision, speech recognition, language translation, and tasks of understanding natural language [6–11].

A vast amount of areas in research, engineering, and other aspects of human endeavor are now influenced by the developments of machine learning over the past decade, and this impact is only going to grow. Combined with the stagnation of

DOI: 10.1201/9781003161233-9

general-purpose CPU efficiency enhancements in the post-Moore Law period, the advanced computing requirements of machine learning mark an exciting moment for the computer hardware market [11–14]. Scientific journals provide a range of knowledge that can be valuable for analysis, decision-making in management, effect evaluation, etc. But these data are often stuck behind the PDF standard now. While there are methods for extracting text and other details from PDF documents, the resultant production frequently falls short of analysis requirements. Any of the features that trigger problems for conventional PDF extraction methods, as well as tools for analyzing extracted information created on top of these methods, include captions, statistics, charts, header, and footer details. Finally, it gives a sketch of at least one promising path towards multi-task models of far larger size that are sparsely enabled and use far more complex, example- and task-based routing than today's machine learning models [15–18].

9.2   DEEP LEARNING

Deep Learning models have a common layout, resembling the brain's pattern detection units, of individual neurons. These neurons begin to understand characteristics by the tuning of their weights in the training period. It is evaluated on unseen data after the neurons are set and the model does well on the training set, and now the neurons are responsible for detecting their respective trends on the test set. If a neuron finds the template, the data is transmitted to the next layer of neurons that understand more complicated characteristics. The usage of machine learning to learn to automatically produce high quality solutions for a variety of different Nondeterministic Polynomial (NP)-hard optimization issues that occur in the overall workflow for the design of custom Application Specific Integrated Circuits (ASICs) is one field that has considerable potential. When the final design of an ASIC is fleshed out, present placement and routing for complicated ASIC designs involves vast teams of human placement specialists to iteratively refine from high-level placement to specific placement. Since there is substantial human intervention in the positioning phase, it is inconceivable to suggest drastically new formats after the original high level design is finished without significantly impacting the timeline of a chip project. Our current findings demonstrate that the distinction between the body text and other parts of a PDF document can be effectively distinguished and trained by a deep learning network. The next move is to expand the method to describe each form of text (title, speaker, abstract, body text, etc.) rather than only body text versus other text. In addition, by introducing more info, we expect to improve the precision of our network and to build an extraction method that leverages the deep learning network's performance to extract text. Although we are currently testing accuracy on the basis of an inferred versus redacted image per pixel count, an enhanced accuracy measure will be to leverage such an extraction technique to identify the accuracy of this text extraction method per character.

In brief, deep learning has become a recent frontier of machine learning and has attracted wide attention in numerous areas of study. In certain areas, it has seen benefits over conventional approaches of machine learning. While deep learning works well in certain tasks of machine learning, in other ways it works just as poorly as the other techniques of learning. In addition to other deep learning research, there is a need to create scientific, sound theoretical foundations for deep learning. With some progress, deep learning has been applied to natural language processing. The outcome of deep learning appears to be encouraging, but the observations are tentative from several NLP subfields, only from a few study groups. In comparison, for NLP, the outcome is also far from satisfactory, enabling machines to comprehend human languages. For both deep learning and NLP, further investigations are required.

1. Bayesian Networks
2. Markov Chains
3. Back Propagation
4. Reinforment Learning
5. Perception
6. CNN
7. RNN

Deep Learning is a new field of research in Machine Learning that has been introduced with the goal of bringing Machine Learning closer to one of its original objectives: artificial intelligence. Deep Learning is about learning multiple levels of representation and abstraction that enable information such as images, sound, and text to make sense. The current fields of research such as neural networks, graphical models, feature learning, unsupervised learning, optimization, pattern recognition, and signal processing have been interacted with and closely related to deep learning. This is also driven by neuroscience, more similarities to our brain intelligence understanding (learning from unlabeled data), and there are already numerous applications in computer vision, speech recognition, processing of natural language, recognition of hand writing, and so on. For the future tasks involved in machine learning, deep learning is one of the progressive and promising areas of machine learning, particularly in the area of neural networks.

9.3   DEEP LEARNING NETWORK

While Machine Learning (ML) has become associated with Artificial Intelligence (AI), Deep Learning (DL) has recently been persistently used instead of machine learning. If statistics are grammar and poetry is machine learning, then Socrates' development is deep learning. Though machine learning is interested in controlled and unsupervised approaches, deep learning continues the motivation to mimic the human nervous system by integrating advanced forms of neural networks (NN). Deep learning finds its uses in numerous AI solutions because of its practicality, such as machine vision, natural language processing, intelligent visual analytics, hyperspectral imagery detection from satellites, and so on. We tried to show good learning capacity and better use of the dataset by deep learning for feature extraction. With its past, development, and introduction to some of the advanced neural networks such as Convolution Neural Network (CNN) and Recurrent Neural Network (RNN), this paper gives an introductory tutorial to the realm of deep learning. This thesis would serve as an introduction to the incredible area of deep learning and its possible application in working with today's vast chunks of unstructured data, which may take decades for people to grasp and retrieve useful knowledge.

The capacity to learn, which is achievable by machine learning, is a significant move in our mission to render our computers intelligent. It requires learning that is controlled, semisupervised, unsupervised and reinforcing. For a decade, algorithmic methods flourished from neural networks before early machine learning. Deep learning derives its influence from the human nervous system, which has multi-layer neural networks that are hierarchically related. It is an evolving area in machine learning that, owing to its early successes in computer vision and image recognition, has gained popularity. There are several distinct layer-by-layer similarities between these artificial neural networks and the path of data propagation.

Unlike more traditional machine learning and function engineering algorithms, Deep Learning has the benefit of delivering a way to solve the challenges of data processing and learning contained in large input data volumes. More importantly, it helps to derive dynamic data representations automatically from vast quantities of unsupervised data. This makes it a powerful method for Big Data Analytics, which requires data mining that is typically unsupervised and uncategorized from very broad sets of raw data. In Deep Learning, the hierarchical learning and extraction of various levels of complex data abstractions offers a certain degree of simplification for Large Data Analytics activities, especially for the analysis of large data volumes, textual indexing, data labeling, data retrieval, and selective tasks such as classification and prediction.

Machine learning is a sub-area of artificial intelligence that empowers automated applications without being explicitly designed to get into a state of self-learning. These frameworks are encouraged to develop, improve, evolve and adopt on their own when faced with new data. Machine Learning focuses on the development of programmers that can navigate and use knowledge on their own. The details and patterns are detected through Machine Learning. In reviving interest in deep learning, unsupervised learning has had a catalytic impact, but has since been overshadowed by the advances of solely supervised learning. Although we have not concentrated on it in this analysis, in the longer term, we anticipate unsupervised learning to become even more relevant. Human and animal learning is essentially unsupervised: through studying it, we discover the nature of the universe, not by being informed every object's name. Deep learning is becoming a mainstream technology for industrial-scale speech recognition. To demonstrate and analyze the strengths and weaknesses of the techniques described in the paper, selected experimental results, including speech recognition and related applications such as spoken dialogue and language modeling, are presented. Potential improvements to these methods and future directions for research are discussed.

Overseen learning is the most common form of machine learning, deep or not. Imagine that we want a system to be built that can classify images as, say, a house, a car, a person or a pet. We first collect a large data set of images, each labeled with its category, of houses, cars, people and pets. The machine is shown an image during training and produces an output in the form of a vector of scores, one for each category. Of all categories, we want the desired category to have the highest score, but before training, this is unlikely to happen. We calculate an objective function that measures the error between the output scores and the desired pattern of scores (or distance). To decrease this error, the machine then modifies its internal adjustable parameters. These adjustable parameters, often referred to as weights, are true numbers that can be seen as 'knobs' that define the machine's input-output function. There may be hundreds of millions of these adjustable weights in a typical deeplearning system, and hundreds of millions of labeled examples to train the machine with. The learning algorithm computes a gradient vector to properly adjust the weight vector that indicates by what amount the error would increase or decrease if the weight was increased by a small amount for each weight. The weight vector is then adapted to the gradient vector in the opposite direction. Multiple arrays, such as a colour image consisting of three 2D arrays containing three colour channel pixel intensities. Several data modalities take the form of multiple arrays: 1D, including language, for signals and sequences; 2D, for images or audio spectrograms; and 3D, for video or volumetric images. ConvNets has four key ideas that take advantage of natural signal properties: local connections, shared weights, pooling and the use of many layers.

Two types of layers are composed of the first few phases: convolution layers and pooling layers. Units are organized into function maps in a convolution layer, within which each unit is linked through a set of weights called a filter bank, to local patches in the function maps of the previous layer. A non-linearity such as a ReLU is then passed through the outcome of this local weighted sum. The same filter bank is shared by all units in a feature map. Different maps of features in a layer use various filter banks. Although the convolutionary layer's role is to detect local feature conjoins from the previous layer, the pooling layer's role is to merge similar features into one semantically. Because the relative positions of the characteristics forming a motif can vary somewhat, it is possible to detect the motif reliably by coarsegraining the position of each function. In one feature map, a typical pooling unit calculates the maximum of a local patch of units (or in a few feature maps). Input from patches moved by more than one row or column is taken by neighboring pooling units, thereby reducing the dimension of the representation, and creating an invariance to small shifts and disruptions. Non-linearity and pooling are stacked, followed by more convolutionary and fully-connected layers, two or three stages of convolution. Back propagating gradients through a ConvNet is as simple as through a regular deep network, allowing all the weights in all the filter banks to be trained.

9.4 DEEP LEARNING NEURAL NETWORK

The output of the first layer is correlated with the input of the next layer of artificial neural networks and so on before the final layer. Neural networks are trained using a backpropagation algorithm, which conducts the reverse gradient calculation from the last layer to the first layer. It will require a variety of signals to combine and once the meaning reaches a given level, the Perceptron was the first mechanical equivalent to the human brain neuron.

Input Layer – –– → Data – – – – → Feature Extraction – – →
Combined data – – → Feature Extraction – –– → Data – – – –– → Output Layer

Like the ANNs of nodes (neurons), CNNs utilize kernels], which are essentially neuron grids that can learn patterns in images. CNNs are a particular form of NN developed for image recognition. Kernels learn simple features including horizontal and vertical lines in the initial layers of CNN. The kernel continues to recognize complicated features such as face, nose, lips, etc. as we go further through the layers, and the network may define the whole picture centered on this CNN network in the final layer.

The deep learning area deals with several algorithms, each programmed to solve a particular problem. Any of the algorithms mentioned below are

- Feed Forward Neural Networks (Artificial Neuron).
- Radial Basis Function Neural Network.
- Multilayer Perception.
- Unsupervised pretrained network
  - Auto encoders
  - Deep Belief Networks
  - Generative Adversarial Networks (GANs)
- Convolution Neural Networks
- Recurrent Neural Network/LSTMs
- Recursive Neural Networks

CNNs are a class of multi-layer neural networks that are primarily developed for the use of images and videos on two-dimensional data. Earlier work in time-delay neural networks (TDNN) affects CNNs, which decrease the criteria for learning computation by exchanging weights in a temporal dimension and are intended for processing speech and time series. The first genuinely efficient deep learning technique is CNNs, where multiple levels of a hierarchy are efficiently educated in a rigorous way. A CNN is a topology or design option that leverages spatial relationships to minimize the number of parameters to be trained and hence facilitates the general training of feed-forward back propagation. As a deep learning system that is driven by limited criteria for data preprocessing, CNNs were suggested. In CNNs, small portions of the picture are viewed as inputs to the lowest layer of the hierarchical system (called a local receptive field). In general, knowledge propagates across the multiple layers of the network, whereby digital filtering is applied to each layer to extract salient features of the data detected. As the local receptive area gives the neuron or processing unit access to elementary features such as aligned edges or corners, the approach offers a degree of invariance to move, scale and rotate.

The secret units are learned to capture the associations of higher-order data found in the visible units. Initially, the layers of a DBN are linked only by guided top-down generative weights, apart from the top two layers, which create an associative memory. RBMs are desirable as a building block, owing to their simplicity of understanding these link weights, over more conventional and heavily layered sigmoid belief networks. The initial pre-training takes place in an unsupervised, selfish layer-by-layer way to achieve generative weights. Deep neural networks exploit the property that compositional hierarchies are many natural signals in which higher-level characteristics are obtained by composing lower-level ones. Local edge combinations form motifs in images, motifs are assembled into parts, and objects are formed by parts. Similar hierarchies, from sounds to phones, phonemes, syllables, words and phrases, exist in speech and text. When elements in the previous layer vary in position and appearance, the pooling enables representations to differ very little. In reviving interest in deep learning, unsupervised learning had a catalytic impact, but has since been overshadowed by the achievements of purely supervised learning. While we have not concentrated on it in this review, in the longer term, we expect unsupervised learning to become far more important. Human and animal learning is largely unsupervised: by observing it, we discover the structure of the world, not by being told every object's name.

Human vision is an active process that sequentially samples the optical array using a small, high-resolution fovea with a large, low-resolution surrounding in an intelligent, task-specific way. We expect that much of the future vision advancement will come from end-to-end trained systems that combine ConvNets with RNNs that use reinforcement learning to decide where to look. Systems that combine deep learning and reinforcement learning are in their infancy, but in classification tasks they already outperform passive vision systems99 and produce impressive results in learning to play many different video games. Another area in which deep learning is likely to have a big impact over the next few years is natural language understanding. Ultimately, through systems that combine representation learning with complex reasoning, major advances in artificial intelligence will come about. While deep learning and simple reasoning have long been used for the recognition of speech and handwriting, new paradigms are needed to replace rule-based manipulation of symbolic expressions by large vector operations. Image classification's fundamental task is to ensure that all images are categorized according to their specific sectors or groups. Classification is simple for humans, but machines have proven to have major problems. Compared to detecting an object, it consists of unidentified patterns, as it should be classified into the correct categories. Using image classification technology, different applications such as vehicle navigation, robot navigation, and remote sensing are used. Challenging work is still underway and limited resources are required to improve it.

Image classification in machine vision has become a major challenge and has a long history. A broad intra-class range of images caused by colour, size, environmental conditions and shape are included in the challenge. Big Data from labeled training images are needed and it takes a lot of time and cost to prepare these Big Data for the training purpose alone.

With Python as the programming language for image classification, deep neural network, based on Tensor Flow, is used in this paper. In this project, thousands of images are used as the input data. The accuracy of each 'train' session percentage will be studied and compared. The models used in natural language processing (NLP) for machine learning (ML) are often applicable to other fields. The idea of a pipeline to process raw data information to solve a useful problem is often constructed in a similar way, irrespective of the domain to which the ML is applied. Techniques and the construction of an ML pipeline could therefore be transferred to other problems, such as the processing of data corresponding to patients with heart transplants. A large proportion of image retrieval queries relate to individuals and objects, the 'story' within the image, as well as relationships between them. Although the automatic recognition, detection, and segmentation of objects in images has reached remarkable levels of precision, it is still a territory that is still largely unexplored to identify relationships. However, the identification of these relationships would allow users to search more accurately for images illustrating two or more objects. Relationships within images between objects are often ambiguous and captions are meant to assist us in their interpretation. We often have to read the caption or the surrounding text as human beings to understand what occurred and the nature of the relationships between the entities. In automatic interpretation, this combined use of text and images has been explored.

It is important to note that the changes in the deep architecture layers are non-linear transformations that attempt to extract the data's underlying explanatory factors. As transformation algorithms in the layers of the deep structure, a linear transformation such as PCA can not be used because the compositions of linear transformations produce another linear transformation. There would therefore be no point in having an architecture that is deep. For example, by providing the Deep Learning algorithm with some face images, the edges can be learned in different orientations on the first layer; these edges are composed in the second layer to learn more complex characteristics such as different parts of a face such as lips, nose and eyes. It composes these features in the third layer to learn even more complex characteristics, such as different people's face shapes. These final representations can be used as features in face recognition applications. This example is provided to simply explain how a deep learning algorithm finds more abstract and complex representations of data by composing representations acquired in a hierarchical architecture in an understandable way. It should, however, be taken into account that deep learning algorithms do not necessarily attempt to construct a pre-defined sequence of representations on each layer (such as edges, eyes, faces), but instead perform non-linear transformations on different layers more generally. Such transformations tend to disentangle data variation factors. It is still one of the main open questions in deep learning algorithms to translate this concept to appropriate training criteria.

The final representation of the deep learning algorithm data (output of the final layer) provides useful information from the data that can be used as features in building classifiers or can even be used for data indexing and other applications that are more efficient than high dimensional sensory data when using abstract data representations. A difficult optimization task, such as learning the parameters in neural networks with many hidden layers, is to learn the parameters in a deep architecture. The sensory data is fed to the first layer as learning data at the start. Based on this data, the first layer is then trained, and the output of the first layer (the first level of learned representations) is supplied to the second layer as learning data. Until the desired number of layers is obtained, such iteration is done. The Deep Network is being trained at this point. For various tasks, the representations learned in the last layer can be used. If the task is a classification task, a supervised layer is usually placed on top of the last layer and its parameters are learned (either randomly or by using supervised data or keeping the rest of the network fixed). The entire network is fine-tuned at the end by providing it with supervised data.

Deep Learning algorithms, as previously stated, extract meaningful abstract representations of raw data by using a hierarchical multi-level learning approach, where more abstract and complex representations are learned at a higher level based on the less abstract concepts and representations at the lower level(s) of the hierarchy of learning. While Deep Learning can be used if it is available in sufficiently large quantities to learn from labeled data, it is primarily attractive for learning from large quantities of unlabeled/unsupervised data, making it attractive for Big Data to extract meaningful representations and patterns. Once the abstractions of hierarchical data are learned with Deep Learning from unsupervised data, more conventional discriminatory models can be trained with the help of relatively fewer supervised/labeled data points, where the labeled data is usually obtained by human/expert input. Compared to relatively shallow learning architectures, deep learning algorithms are shown to perform better at extracting non-local and global relationships and patterns in the data.

The capacity to learn, which is achievable by machine learning, is a significant move in our mission to render our computers intelligent. It requires learning that is controlled, semisupervised, unsupervised and reinforcing. For a decade, algorithmic methods flourished from neural networks before early machine learning. Deep learning derives its influence from the human nervous system, which has multi-layer neural networks interconnected hierarchically. It is an evolving area in machine learning that, owing to its early successes in computer vision and image recognition, has gained popularity. There are several distinct layer-by-layer similarities between these artificial neural networks and the path of data propagation. The output of the first layer of artificial neural networks is related to the input of the next layer and so on before the final layer is generated and the outcome is produced. If machine learning is an artificial intelligence sub-discipline, so deep learning may be considered a machine learning sub-discipline. Depending on the application, there are distinct algorithms, which are listed as defined in Table 9.1. Supervised learning requires training a neural network with labeled datasets and also defining the target outcome, which can be used in order to predict outcomes of related data types, but the challenge lies in accessing labeled data sets that are both small in number and costly. Semisupervised learning utilizes neural networks equipped from a combination of data sets that are both classified and unlabeled. When practicing the inputs, the machine is granted inadequate data visibility, and some of the performance goals in the network are absent. Comparison of publication from 2000–2020 is shown in Figure 9.1.

A brief Automatic Speech Recognition survey presents and discusses the main themes and advances made in research over the past 60 years, in order to provide a technological perspective and to appreciate the fundamental progress made in this important field of speech communication. Speech recognition system design requires careful attention to the following issues: definition of different types of speech classes, speech representation, techniques of feature extraction, speech classifiers, and database and performance assessment. The aim of this review paper is to summarize and compare some of the well-known techniques used in different phases of the system of speech recognition and to identify research topics and applications at the forefront of this exciting and challenging field. Speech is the human being's most prominent and main communication mode. Human computer interaction is known as

TABLE. 9.1
Number of Papers Published in Deep Learning Network

Year	Number of Papers Published					
	IEEE	IET	ELSEVIER	SPRINGER	WILEY	ACM
2000	1	2	2	3	1	1
2001	2	6	5	2	5	2
2002	5	8	9	6	8	5
2003	8	4	8	7	10	8
2004	6	8	7	2	12	6
2005	5	10	10	5	15	16
2006	8	15	12	7	12	18
2007	6	17	8	8	14	19
2008	5	18	5	3	16	22
2009	5	20	16	10	18	16
2010	8	22	18	12	12	10
2011	9	26	19	16	15	12
2012	5	18	22	6	16	16
2013	7	21	24	18	18	6
2014	8	22	26	25	20	18
2015	10	18	28	29	10	14
2016	15	12	29	30	20	16
2017	20	18	30	35	20	18
2018	22	20	22	34	10	12
2019	24	18	28	38	10	15
2020	26	22	22	40	20	16

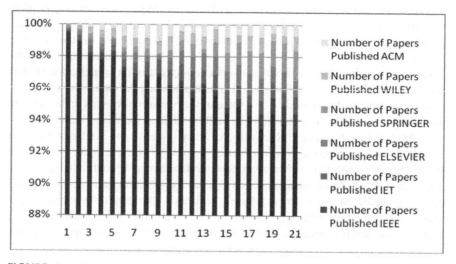

FIGURE. 9.1   Comparison chart publication in 2000–2020.

the human computer interface. The major technological perspective and appreciation of the fundamental advancement of speech recognition also provides an overview of the technique developed at each stage of speech recognition. Speech has the potential to be an important mode of interaction with computers. As per stage, a comparative study of different techniques is done. This paper concludes with the decision on the direction of features for developing techniques using Marathi Language in the human computer interface system. Speech has evolved as a primary form of human communication. The advent of digital technology has provided us with highly versatile, high-speed, low-cost and high-power digital processors that allow researchers to convert analog speech signals into digital speech signals that can be scientifically studied. The main considerations for developing an efficient Automatic Speech Recognition system are to achieve greater recognition accuracy, low word error rate and address the issues of variability sources. Feature extraction requires a lot of attention in speech recognition, because recognition performance relies heavily on this phase. An effort has been made in this paper to highlight the progress made so far in the feature extraction stage of the speech recognition system and to discuss an overview of the technological perspective of the Automatic Speech Recognition system.

The precision of speech recognition has been dramatically improved by deep learning systems, and in recent years, various deep architectures and learning methods have been developed with distinct strengths and weaknesses. The focus of this paper is how ensemble learning can be applied to these varying deep learning systems to achieve greater accuracy in recognition. With applications specifically to speech-class posterior probabilities as computed by convolutive, recurrent, and fully-connected deep neural networks, we develop and report linear and log-linear stacking methods for ensemble learning. Problems of convex optimization are formulated and solved, with analytical formulas derived for training the parameters of ensemble-learning. After stacking the deep learning subsystems that use various mechanisms for computing high-level, hierarchical features from the raw acoustic signals in speech, experimental results demonstrate a significant increase in phone recognition accuracy.

9.5 APPLICATIONS OF DEEP LEARNING

Machine learning is about learning to change later in the future, based on what was already learned. The aim is to devise learning algorithms without human assistance or guidance that do the learning naturally. Speech recognition is the simplest and most powerful way to interact in such a way that contact is interchangeable. This is why researchers are motivated to create a computer interface optimized for speech recognition. We will interact even more with our computers this way than with the conventional mouse and keyboard, joystick, etc. But a massive, dynamic phenomenon takes up development and recognition of voice/speech by machines. Automatic speech recognition (ASR) is currently driven by deep learning neural networks such as voice search/speech transcription for the application. As it uses advanced equipment for sorting, ASR is a complicated and cumbersome task. Object detection is known to be a programming process that is not trivial. For several machine learning algorithms, the MNIST digit picture classification problem has been used as a benchmark. One of the early uses of neural networks, especially convolution (or time-delay) neural networks, was speech recognition. In the field of speech recognition, the recent resurgence of interest in neural networks, deep learning, and representation learning has had a strong impact. It will be feasible to train such a hierarchical network on a wide range of observations, assuming robust deep learning is accomplished, and later extract signals from this network to a reasonably simple classification engine for the purpose of robust pattern recognition. Robustness here refers to the capacity to demonstrate invariance in classification to a number of transformations and distortions, including noise, scale, rotation, different lighting situations, displacement, etc.

Big Data is an emerging concept that defines any quantity of structural, semi-structured and unstructured data that has the potential to be exploited for information. Big Data is comparatively modern, the means to capture and preserve enormous data measures for inevitable study. Big Data is a major dataset and a type of computational techniques and technology used to manage these massive datasets. Big Data is the data of the largest variation that exists with greater velocity in expanding volumes. An automated learning algorithm can be challenging to process data for more heterogeneous, under-represented groups and noiseimpaired. Although the strategies previously described, such as transfer learning with highperformance learners, can efficiently acquire data from multiple domains, they are selfish and sluggish. It is also a dynamic and non-productive one to pre-process it owing to the substantial scale of the data attribute. Big Data often receives a wide variety of sources, data sets of high dimensionalities, and further results. In coping with these resources, deep learning mechanisms with a central processor and storage pose a difficulty.

Advanced approaches involve many mechanisms through which Deep Learning and Large Data has the ability to solve the challenges of Machine Learning. Despite having little obstacles, deep learning has the potential to interact with and learn issues contained in vast amounts of input data. Overseen learning is the most popular method of machine learning, deep or not. Imagine that we want a machine to be developed that can identify photos as, say, a home, a vehicle, a human or a pet. We first compile a broad data collection of photos, each labelled with its category, of buildings, vehicles, people and pets. The computer is shown a picture during training and generates a performance in the form of a vector of scores, one for each rank. Among all divisions, we expect the optimal group to get the maximum ranking, but before practicing, this is impossible to happen. We quantify an analytical feature that calculates the error between the performance scores and the target pattern of scores (or distance). To decrease this malfunction, the computer then modifies its internal adjustable parameters. These adjustable parameters, also referred to as weights, are true numbers that can be seen as 'knobs' that determine the machine's input-output operation. There could be hundreds of millions of these customizable weights in a standard deep-learning framework, and hundreds of millions of named examples to practise the computer with. The learning algorithm computes a gradient vector to better change the weight vector that shows by what extent the error will increase or decrease if the weight was raised by a small amount for each weight. The weight vector is then adapted to the gradient vector in the reverse direction.

$$\text{Training Data Set} - \!-\!- \rightarrow \text{Convolution Layer} - \!-\!- \rightarrow$$
$$\text{Up sampling} - \!-\!- \rightarrow \text{Connected Layer} - \!-\!-\!- \rightarrow \text{Output}$$

Human vision is an active mechanism that sequentially samples the optical array utilizing a tiny, high-resolution fovea with a large, low-resolution surrounding in an insightful, task specific manner. We believe that most of the potential vision advancement will come from end-to-end qualified systems that mix ConvNets with RNNs that use reinforcement learning to determine where to look. Another field in which deep learning is going to have a major effect within the next few years is natural language comprehension. As they learn techniques for narrowly listening to one component at a time, we believe programmers that use RNNs to comprehend sentences or whole documents will get even easier. There are their own drawbacks on all data-parallel and model-parallel methods. On one hand, if there are so many training modules for data parallelism, the learning rate needs to be lowered to render the training procedure smooth. On the other hand, if there are so many segmentations in model parallelism, the performance from the nodes can increase dramatically and lower the performance accordingly. Generally speaking, the bigger the dataset is, the more desirable it is to provide data parallelism. The broader the model in deep learning, the more important it is to include parallelism in the model. In comparison, the communication necessary for synchronization in model parallelism is difficult to mask relative to data parallelism because only partial information is used for the entire batch in each node, although some advanced frameworks such as Tensor Flow support asynchronous kernels to save the cost of communication. Therefore, before going on to the next layer, it is important to wait until the synchronization phase completes, as the events may not be processed with just partial details.

$$\text{Data Set} - \!-\!-\!- \rightarrow \text{Algorithm} - \!-\!- \rightarrow \text{Output}$$

In the computer vision group, video analytics has drawn significant interest and is regarded as a difficult challenge as it requires both spatial and temporal knowledge.

Audio processing is the mechanism that acts on electric or analog audio signals directly. Speech recognition (or speech transcription), speech amplification, telephone classification and music classification are important. Due to its relevance in perfect human-computer interaction, speech processing is an active research field.

9.6   CONCLUSION

Centered on a mixture of convolution and recurrent neural networks, a modern paradigm. The configuration of the tree, allows several vectors to be combined, use several RNN weights and keep parameters randomly initialized, unlike previous RNN models. This allows for fast speeds and parallelization, outperforms two layer CNNs and obtains state-of-the-art output without any external characteristics. Demonstrate the applicability of the modern depth picture domain of convolution and recurrent function learning. This technique can be seen as a semi-monitored learning process, in which the labeled data are inadequate to train a deep network as a whole. High-speed data of time shifting distributions is another problem that piles up on the topic of online learning. The retail and banking data pipelines that carry enormous market principles reflect this threat.

REFERENCES

[1] V. Ferrari, M. Marin-Jimenez, and A. Zisserman, 2008, *Progressive search space reduction for human pose estimation, IEEE Conference on Computer Vision and Pattern Recognition* 10.1109/CVPR.2008.4587468

[2] M. A. Fischler, and R. A. Elschlager, The representation and matching of pictorial structures. *Computers*, IEEE Transactions on Computers, 100, 1: 67–92, 1973.

[3] R. Girshick, J. Donahue, T. Darrell, and J. Malik, 2014, *Rich feature hierarchies for accurate object detection and semantic segmentation, 2014 IEEE Conference on Computer Vision and Pattern Recognition*, 10.1109/CVPR.2014.81

[4] G. Gkioxari, P. Arbeláez, L. Bourdev, and J. Malik, 2013, *Articulated pose estimation using discriminative armlet classifiers, 2013 IEEE Conference on Computer Vision and Pattern Recognition*, 10.1109/CVPR.2013.429

[5] Marc'Aurelio Ranzato, Volodymyr Mnih, Joshua M. Susskind, and Geoffrey E. Hinton, 2013. Modeling natural images using gated MRFs. *IEEE Transactions on Pattern Analysis and Machine Intelligence*, 35, 9: 2206–2222, 2013.

[6] Alessio Micheli, Neural network for graphs: A contextual constructive approach. *IEEE Transactions on Neural Networks*, 20, 3: 498–511, 2009.

[7] Samira Pouyanfar, and Shu-Ching Chen, Automatic video event detection for imbalance data using enhanced ensemble deep learning, *International Journal of Semantic Computing*, 11, 1, 85–109, 2017.

[8] Samira Pouyanfar, and Shu-Ching Chen, 2017, *T-LRA: Trend-based learning rate annealing for deep neural networks, IEEE International Conference on Multimedia Big Data.* IEEE, 50–57.

[9] Samira Pouyanfar, Shu-Ching Chen, and Mei-Ling Shyu, 2017, *An efficient deep residual-inception network for multimedia classification, International Conference on Multimedia and Expo.* IEEE, 373–378.

[10] Antonio Torralba, Rob Fergus, and William T. Freeman, 80 million tiny images: A large data set for nonparametric object and scene recognition. *IEEE Transactions on Pattern Analysis and Machine Intelligence*, 30, 11: 1958–1970, (2008).

[11] S. Salehi, A. Selamat, and M. Bostanian, 2011, *Enhanced genetic algorithm for spam detection in email, 2011 IEEE 2nd International Conference on Software Engineering and Service Science*, Beijing, pp. 594–597.

[12] A. Chakrabarty, and S. Roy, 2014, *"An optimized k-NN classifier based on minimum spanning tree for email filtering," 2014 2nd International Conference on Business and Information Management (ICBIM)*, Durgapur, pp. 47–52.

[13] Chharia and R. K. Gupta, 2013, *Email classifier: An ensemble using probability and rules, 2013 Sixth International Conference on Contemporary Computing (IC3)*, Noida, pp. 130–136.

[14] T. Vyas, P. Prajapati, and S. Gadhwal, *2015, A survey and evaluation of supervised machine learning techniques for spam e-mail filtering, 2015 IEEE International Conference on Electrical, Computer and Communication Technologies (ICECCT)*, Coimbatore, pp. 1–7.

[15] Du Tran, Lubomir Bourdev, Rob Fergus, Lorenzo Torresani, and Manohar Paluri, 2016, *Learning spatiotemporal features with 3D convolutional networks*, *IEEE International Conference on Computer Vision*. IEEE, 4489–4497.

[16] Shiliang Sun, Changshui Zhang, and Guoqiang Yu, A Bayesian network approach to traffic flow forecasting. *IEEE Transactions on Intelligent Transportation Systems*, 7, 1: 124–132, 2006.

[17] Jie Cheng, Russell Greiner, Jonathan Kelly, David Bell, and Weiru Liu, Learning Bayesian networks from data: An information-Theory based approach. *The Artificial Intelligence Journal*, 137: 43–90, 2002.

[18] Ramon Lopez De Mantaras, and Eva Armengol, Machine learning from examples: Inductive and Lazy methods. *Data and Knowledge Engineering*, 25, 1–2: 99–123, 1998.

Index